Genetic manipulation: impact on man and society

Genetic manipulation: impact on man and society

Edited by
WERNER ARBER
KARL ILLMENSEE
W. JAMES PEACOCK
PETER STARLINGER

Technical editor: A. Padron

Published on behalf of
The ICSU Press by
Cambridge University Press
Cambridge
London New York New Rochelle
Melbourne Sydney

Published by the Press Syndicate of the University of Cambridge
The Pitt Building, Trumpington Street, Cambridge CB2 1RP
32 East 57th Street, New York, NY 10022, USA
296 Beaconsfield Parade, Middle Park, Melbourne 3206, Australia

First published 1984

Library of Congress catalogue card number: 83-26166

British Library Cataloguing in Publication Data

Genetic manipulation
    1. Genetic engineering
    I. Arber, Werner
    575   QH442
ISBN 0 521 26417 0

Transferred to digital printing 2004

Contents

MICRO-ORGANISMS

GENETIC MANIPULATION OF THE MAMMALIAN GERM LINE

Speakers and Editors

ARBER, W. - Department of Microbiology, Biozentrum of the University of Basel, Klingelbergstrasse 70, CH-4056 Basel, Switzerland

BERNARDI, G. - Laboratoire de Genetique Moleculaire, Institut Jacques Monod, 2 Place Jussieu, 75005 Paris, France

CHAKRABARTY, A.M. - Department of Microbiology, University of Illinois Medical Center, 835 S. Wolcott, Chicago, IL 60612, USA

CROCE, C.M. - Wistar Institute of Anatomy and Biology, Philadelphia, PA 19104, USA

DAY, P.R. - Plant Breeding Institute, Maris Lane, Trumpington, Cambridge CB2 2LQ, UK

EHRLICH, S.D. - Institut de Recherche en Biologie Moleculaire, Universite Paris VII, 2 Place Jussieu, 75221 Paris, France

FRANKEL, O. - Division of Plant Industry, CSIRO, Box 1600, Canberra City, ACT 2601, Australia

GOEBEL, W. - Institut fur Genetik und Mikrobiologie, Rontgenring 11, 87 Wurzburg, Federal Republic of Germany

HUNLICH, T.L.A. - Universitats Fraunenklinik, Universitatstrasse 21/23, D-8520 Erlangen, Federal Republic of Germany

ILLMENSEE, K. - Laboratory of Cell Differentiation, University of Geneva, 20 rue Ecole Medecine, CH-1211 Geneve 4, Switzerland

JAENISCH, R. - Heinrich-Pette-Institut fur Experimentelle Virologie und Immunologie an der Universitat Hamburg, Martinstrasse 52, 2000 Hamburg 20, Federal Republic of Germany

LEVY, S.B. - Department of Medicine and of Molecular Biology and Microbiology, Tufts University School of Medicine, 136 Harrison Avenue, Boston, MA 02111, USA

LORZ, H. - Max Planck Institut, D-5000 Koln 30, Federal Republic of Germany

LUTHY, P. - Institut fur Microbiologie, ETH-Zentrum, Universitatstrasse 16, CH-8092 Zurich, Switzerland

PADRON, A.B. - Papanicolaou Cancer Research Institute, P.O. Box 016190, Miami, FL 33101, USA

PALMITER, R.D. - Howard Hughes Medical Institute Laboratory, Department of Biochemistry SJ 70, University of Washington, Seattle, WA 98195, USA

PEACOCK, W.J. - CSIRO Division of Plant Industry, P.O. Box 1600, Canberra City, A.C.T. 2601, Australia

RIEDEL, G. - Genetics Institute, 225 Longwood Avenue, Boston, MA 02115, USA

RIGBY, P.W.J. - Department of Biochemistry, Imperial College of Science and Technology, London SW7 2AZ, England

SCHELL, J. - Laboratory for Genetics, Faculty of Science, Royal University of Ghent, Ledenganckstraat 35, B-9000 Ghent, Belgium and Max Planck-Institut fur Zuchtungsforschung, Koln, Federal Republic of Germany

SPRADLING, A.C. - Department of Embryology, Carnegie Institute of Washington, 115 West University Parkway, Baltimore, MD 21210, USA

STARLINGER, P. - Institute of Genetics, University of Cologne, Weyertal 121, 5000 Cologne-Lindenthal, Federal Republic of Germany

STRANZINGER, G. - Institute for Animal Production, Animal Breeding Section, ETH Zuerich Zentrum, CH-8092 Zurich, Switzerland.

WENZEL, G. - Institute for Resistance Genetics, Grunbach, Federal Republic of Germany

WILLIAMS, P. - International Board for Plant Genetic Resources/AGPG, FAO, 00100, Rome, Italy

Preface

The Third COGENE Symposium "Genetic Manipulation: Impact on Man and Society" was organized with the collaboration of the Institute of Genetics, University of Cologne and of the Federation of European Biochemical Societies by W. Arber, K. Illmensee, W.J. Peacock and P. Starlinger. The Symposium took place during April 7-9, 1983 in Cologne, Federal Republic of Germany.

The aim of the Symposium was to bring together a number of scientists, who gave state-of-the-art presentations of genetic manipulation in the fields of microbiology and of plant and animal molecular biology. The selection of speakers was such that possible future applications of genetic manipulation and their impact on man and society could be discussed. A few of the contributions dealt not with laboratory work, but with broader areas, and discussed future breeding goals and gene erosion in the plant section, as well as animal breeding and in vitro fertilization of human eggs in the animal section. In addition to the formal presentations, two open discussion sections were organized dealing with wider problems including ethical issues. These were met with interest by an audience of 600 participants.

This book includes most, but unfortunately not all, contributions and certainly not all discussions. It is hoped that, even so, it will be of some interest to a wider audience than could be in Cologne.

G. Bernardi
Chairman, COGENE

The Committee on Genetic Experimentation

The Committee on Genetic Experimentation (COGENE) was created by decision of the 16th General Assembly of ICSU and came into being in 1977. It is a Scientific Committee, one of a type formed by ICSU whenever a task is defined that is of major interest to several unions and the program is of a long-term nature. COGENE is supported by 9 member unions - Pure and Applied Chemistry, Biological Sciences, Biochemistry, Physiology, Pure and Applied Biophysics, Nutritional Sciences, Pharmacology, Immunological Societies and Microbiological Societies - and includes scientists prominent in the field of recombinant DNA research. It was the advent of this new branch of science that prompted the formation of COGENE.

Membership

W. Arber (Switzerland)
G.L. Ada (Australia)
A.A. Bayev (USSR)
G. Bernardi (France, Chairman)
S.N. Cohen (USA)
R.F.W. Hardy (USA)
K. Illmensee (Switzerland)

W.F.H.M. Mommaerts (USA)
J.W. Peacock (Australia)
R. Riley (UK)
A.M. Skalka (USA, Vice Chairman)
D.A.T. Southgate (UK)
P. Starlinger (FRG)
E. Wollman (France)

The work of COGENE is also assisted by observers from the FAO, UNESCO and WHO and by national correspondents.

Purposes and Objectives

The terms of reference of COGENE include the following purposes and objectives:

COGENE is a Scientific Committee of ICSU, established to serve as a nongovernmental, interdisciplinary and international council of scientists and as a source of advice for the benefit of governments, inter-governmental agencies, scientific groups, and individuals, concerning recombinant DNA activities.

Among its purposes shall be:

a) to review, evaluate and make available information on the practical and scientific benefits, safeguards, containment facilities and other technical matters,
b) to consider environmental, health-related and other consequences of any disposal of biological agents constructed by recombinant DNA techniques,

c) to foster opportunities for training and international exchange, and

d) to provide a forum through which interested national, regional and other international bodies may communicate.

COGENE shall also consider, if necessary, other related activities which may give rise to public concern.

Current Activities

COGENE originally devoted most of its attention to an evaluation of guidelines for the regulation of recombinant DNA research and experiments designed to assess the conjectured hazards of research with recombinant DNA.

With the lessening of concern about the conjectural hazards, and the relaxation of the guidelines pertaining to recombinant DNA research, COGENE has undertaken the organization of training courses in this research in areas of the world where such training was not available but where it was felt that expertly trained biologists could contribute to the advancement of knowledge and the application of such knowledge in their own countries. Accordingly, training courses have been supported in Brazil (1979), India (1981), Yugoslavia (1981), South Africa (1982), Hong Kong (1983) and Costa Rica (1983).

In parallel, COGENE has organized meetings of scientists, representatives of regulatory agencies, research foundations and other interested parties to discuss the state of the art of the research and the public perception, in line with its mandate to act as a forum and as a source of advice for agencies and persons interested in this branch of research.

The first of these symposia was held in 1979 in conjunction with the Royal Society of London and was published under the title "Recombinant DNA and Genetic Experimentation", edited by J. Morgan and W.J. Whelan and is available from Pergamon Press.

In 1981 a COGENE symposium, organized jointly with the Italian National Research Council, addressed the topic "From Genetic Experimentation to Biotechnology: The Critical Transition". This symposium, edited by W.J. Whelan and S. Black, was published for COGENE by John Wiley & Sons Limited.

The International Council of Scientific Unions

The International Council of Scientific Unions (ICSU) is an international non-governmental scientific organization composed of 20 international scientific unions, together with National Members, Scientific and National Associates. Since its creation in 1931 ICSU has adopted a policy of non-discrimination, affirming the rights of all scientists throughout the world - without regard to race, religion, political philosophy, ethnic origin, citizenship, sex or language - to join in international scientific activities.

The principal objective of ICSU is to encourage international scientific activity for the benefit of mankind. It does this by initiating, designing and coordinating international scientific research projects; the International Geophysical Year and the International Biological Programme are probably the best-known examples. ICSU acts as a focus for the exchange of ideas, the communication of scientific information and the development of standards in methodology, nomenclature, units, etc. The various members of the ICSU family organize in many parts of the world conferences, congresses, symposia, summer schools, and meetings of experts, as well as General Assemblies and other meetings to decide policies and programmes. A wide range of publications is produced, including newsletters, handbooks, proceedings of meetings, congresses and symposia, professional scientific journals, data, standards, etc. ICSU now has its own publishing house, the ICSU Press, which is described on the following page.

Committees or Commissions of ICSU are created to organize programmes in multi- or transdisciplinary fields which are not completely under the aegis of one of the Scientific Unions, such as Antarctic, Oceanic, Space and Water Research, Problems of the Environment and Genetic Experimentation. Activities in areas common to all the Unions such as Teaching of Science, Data, Science and Technology in Developing Countries are also coordinated by ICSU Committees.

ICSU maintains close relations and works in cooperation with a number of international governmental and non-governmental organizations, and in particular UNESCO (with which ICSU has taken the initiative of launching a number of international programmes such as the International Indian Ocean Expedition, the World Science and Technology Information System, International Geological Correlation Project, International Biosciences Networks, etc.) and with the World Meteorological Organization (with which ICSU has taken the initiative in launching the Global Atmospheric Research Programme and the World Climate Research Programme).

The ICSU Press

This COGENE symposium is one of the first publications of the ICSU Press, the publishing house of the International Council of Scientific Unions, created in January 1983 by action of the ICSU Executive Board.

The rationale for the formation of this publishing house was that while the individual members of ICSU family are already heavily engaged in publication of major research journals, reviews, monographs, scientific proceedings and the like, by a variety of mechanisms and with a variety of publishers, the need was felt for a central publishing service within ICSU that would carry out the following functions:

1. Implement the publication policies of the Committee on Publications and Communications, the parent body of the ICSU Press.

2. Initiate new serial publications of original research or review, aimed at the international scientific community.

3. Sponsor the publication of works aimed at disseminating scientific findings on an international basis, the better to inform individuals, governments and intergovernmental agencies of the work of scientists and at the same time heighten the general awareness of the work of ICSU and the ICSU family.

4. Act for the Unions and Committees in any of the following ways:
   a) Advisor on contractual negotiations with publishers,
   b) Advisor on technical aspects of publications,
   c) Expert negotiator with publishers on behalf of a Union or Committee
   d) Acting as the publisher for a Union or Committee.

5. Publish monographs, textbooks, data collections, conference proceedings and service publications for and on behalf of ICSU, its Unions and Committees.

6. Establish a network of publishers and printers to act for the ICSU Press, giving preference to the presses of the learned societies and academies.

7. When self supporting, establish and staff a central office to handle all ICSU Press publications and provide an advisory or managerial service to Unions and Committees.

8. Pay special attention to the publication needs of scientists in developing countries and work with ICSU member bodies and other organizations to serve those needs.

9. Engage in cooperative publishing ventures with appropriate international bodies outside the ICSU family whose aims are consonant with those of ICSU.

The ICSU Press Board of Management

E.S. Ayensu	E.A. Flinn
D.P. Den Os (Secretary/Treasurer)	T.F. Malone
L. Ernster	O.G. Tanberg

W.J. Whelan (Chairman)

The Press welcome proposals for publications from any member of the ICSU family and stands ready to advize on all aspects of scientific publication.

Further information on the ICSU Press may be obtained from the ICSU Secretariat, 51 Boulevard de Montmorency, 75016 Paris, France.

Acknowledgements

COGENE receives its financial support from ICSU which in turn receives substantial assistance for its work from UNESCO.

The meeting of which this symposium is a record was made possible by grants from the Deutsche Forschungsgemeinschaft through Sonderforschungsbereich /4 and the Federation of European Biochemical Societies. The ICSU Committee on Science and Technology in Developing Countries (COSTED) povided two fellowships for participants.

The organizers are also grateful to the following companies, located in the Federal Republic of Germany, for financial support.

Amersham Buchler, Braunschweig
BAYER AG, Leverkusen
Beckman Instruments, Munchen
Behringwerke, Marburg
Boehringer, Tutzing
DuPont, Bad Nauheim
Faust, Koln

FMK, Frechen
Heraus Christ, Osterode
HOECHST AG, Frankfurt/Main
Knoll AG, Ludwigshafen
Madaus, Koln
Schleicher & Schull, Dassel
Zeiss, Oberkochen

In the paper by S. Levy (pages 19-28), Figures 2 and 5 are reproduced from reference 11 by permission of the American Society for Microbiology and Figure 4 from reference 6 by permission of Elsevier/North Holland.

The papers by J. Schell et al. (pages 87-102) and W.J. Peacock et al. (pages 115-125) are reproduced from the Miami Winter Symposium, Vol. 20, 1983, "Advances in Gene Technology: Molecular Genetics of Plants and Animals", by permission of the copyright holder and publisher, Academic Press Inc., New York, USA.

MICRO-ORGANISMS

NATURAL MECHANISMS OF MICROBIAL EVOLUTION

Werner Arber

Dept. of Microbiology, Biozentrum of the University of
Basel, Klingelbergstrasse 70, CH-4056 Basel, Switzerland

INTRODUCTION

In the last ten years recombinant DNA technology has become of
primary importance for molecular genetic studies in fundamental biological research as well as for practical applications making use of
the knowledge accumulated in molecular genetics. Recombinant DNA
technology consists in the splicing of a relatively short DNA segment
of any origin into a well selected vector DNA molecule endowed to
replicate autonomously in a host cell. Often well known laboratory
strains of bacteria such as Escherichia coli or Bacillus subtilis are
used as hosts, while plasmids or viral genomes serve as vectors.
Vectors carrying the in vitro inserted segment of foreign DNA are
called recombinant DNA molecules. These multiply after their introduction into appropriate host cells, producing clones of identical
molecules. These can be harvested and used in molecular genetic
studies. Alternatively cells containing a cloned gene can serve for
the preparation of the product of this gene.

Already at the time when this method was first developed, scientists
tried to evaluate any risks inherent to the technique. Mainly two
types of risks can be defined, both related to the possibility of
escape of the manipulated microorganism from its laboratory culture
vessel. 1) Host cells containing the recombinant DNA molecule could
for some reason or another have become a pathogen, presenting the
risk of infection of the exposed personnel. 2) After either
deliberate or fortuitous release from the contained laboratory
condition, a recombinant DNA molecule might in the long term have
some other detrimental effect on the environment.

Specifically designed guidelines for work with recombinant DNA
molecules should limit these risks to a minimum. This, however,
requires a realistic evaluation of the relatively badly defined
risks. Much help for this evaluation comes from a better
understanding of natural processes affecting the stability and the
exchange of genetic material. Some of these processes promote the
natural recombination between nonhomologous genetic information.
Sometimes such recombination occurs between a chromosomal DNA segment
and a natural vector, a viral genome or a plasmid transferable in
cell to cell contact. These processes are well documented in bacteria
and they allow for horizontal flux of genes from one strain of

microorganisms to another. Sometimes, the transferred genes can later become part of the genome of the infected bacterium. In such processes of horizontal gene exchange we see striking similarities to the laboratory experiments involving recombinant DNA molecules. After all, we should remember that the investigators designing the recombinant DNA technique were inspired by natural processes, such as the spontaneous formation of specialized transducing bacteriophage consisting of a vector DNA with an inserted segment of chromosomal genes of the host.

GENETIC STABILITY AND EVOLUTION

The genetic material of practically all living organisms is quite stable in its transmission from generation to generation. However, one can observe rare, spontaneously occurring mutations. In a clone of growing bacteria, for example, this leads to the formation of subclones with altered properties. Sometimes, such a subclone with a particular trait can become selectively enriched in a specific environmental condition. The particular bacterial strain has then made a step of evolutionary change. A number of different molecular mechanisms to be discussed below in more details contribute to spontaneous mutagenesis. Some act by affecting the DNA with a small, local alteration, such as nucleotide substitution. Others in contrast may bring about a more drastic structural alteration, such as the formation of an extended deletion. In addition to these alterations occurring in the vertical transmission of the genetic material, horizontal transmission of genes from one type of organisms to another can also cause a mutation which often represents a major step in the evolutionary development, e.g. if a bacterial strain acquires a new gene which it did not possess before.

TRANSPOSITION

DNA rearrangements by transposition have been studied intensively in recent years, and their important contributions to spontaneous mutagenesis and thus to the biological evolution became obvious (see Shapiro 1983). Transposition acts both at the level of vertical as well as horizontal gene flux. This shall be explained in somewhat more details before we will come back to compare the relative importance of the various mechanisms contributing to spontaneous mutagenesis.

The transposable genetic elements of prokaryotes are classified into IS elements, transposons and some particular viral genomes (Campbell et al. 1979).

IS elements. IS elements are relatively small and do not contain any known gene with functions unrelated to the transposition processes. A typical regular IS element forming part of a bacterial chromosome is about 700 - 1600 basepairs long. It carries at its ends perfect or nearly perfect inverted repeats, the length of which can vary between about 10 and 40 basepairs. Nucleotide sequencing usually reveals 1 to 3 relatively large reading frames, and in some instances it has been shown that these genes are expressed under certain conditions. The involvement of these proteins in transpositional activities is likely (see Iida et al. 1983).

Replicombination. Transposition of IS elements as well as other transposable genetic elements seems to involve a replicative recombination (replicombination). In simplified terms the IS element interacts with a new "target" site on the DNA by recombination and it undergoes a local duplication independently of the replication of the entire chromosome. As a result of this process, one copy of the IS element is usually found at its previous location in the genome, while the second copy has become inserted at another location either in the same or in another DNA molecule in the same cell. Often this insertion may cause a mutation, e.g. if the insertion occurs inside of a gene or also if the insertion occurs inside of an operon, thereby disrupting its normal functioning (Starlinger and Saedler 1976).

IS mediated DNA rearrangements. This simplified picture of transposition may not reflect precisely the molecular mechanisms of the process. Without going into details of complexities identified with particular transposable elements or of published transposition models, a few general principles should be recalled here, but not without pointing to the likelyhood that different IS elements might apply different mechanisms in their transposition processes. According to some models (see Galas and Chandler 1981), intra-molecular transposition of an IS element might result either in directtransposition of the IS element or in the formation of a deletion or of an inversion, depending on the reassortment of the DNA strands upon replicombination. Following the same mechanisms, intermolecular transposition would lead either to direct transposition or to replicon fusion, also called cointegration. With the element IS1, e.g., all of the processes listed, direct transposition, deletion, inversion and cointegration have been documented to occur at frequencies of a roughly similar order of magnitude. These investigations were mostly done with small replicons. For obvious reasons one may expect chromosomes suffering deletion formation often not to survive. Similarly, inversion or cointegration with another replicon may frequently be lethal or confer a growth disadvantage. In all of these processes, only non-lethal events are of course of direct evolutionary relevance. Some of these processes may represent major steps in a chain of evolutionary events by linking previously unlinked genome segments independently of nucleotide sequence homology.

Transposition frequency. Transposition of an IS element is a rare event. Frequencies are difficult to measure and can only be roughly estimated. Depending on the growth conditions, a regular IS element of E. coli may perhaps transpose with a frequency of about 10^{-6} per generation. The nature of the rate limiting factor remains unknown. There are several lines of evidence that besides the expression of genes located on the IS element itself a number of host functions are also essential for successful transposition.

Occurrence of IS elements. Some studied strains of E. coli Kl2 carry
at least 7 and probably more different IS elements. Usually several
copies of each of these IS elements are found in the chromosome. For
a number of other IS elements usually not resident in E. coli Kl2,
transposition activity has been demonstrated after their transfer
from other bacterial strains. Relatively little is known about the
host range of a given IS element. A more profound knowledge of host
range characteristics of IS elements would be important for the
evaluation of the roles played by transpositional processes in
horizontal evolution.

Target selection. Another important question both with regard to
vertical and horizontal evolution concerns the target selection
criteria. The available data suggest that not all transposable
elements apply the same mechanisms. Indeed, some IS elements such as
IS5 (Engler and van Bree 1981), seem to usually insert into a
consensus sequence. Others, such as IS2 also do not select the
insertion target at random, but prefer particular genome regions for
insertion. However, within these hot regions for IS2 insertion, each
individual insertion is found at a different specific location
(Sengstag and Arber 1983). Several IS elements such as IS1 (Galas et
al. 1980; Meyer et al. 1980) preferentially use A-T rich regions as
target. Although we are far from understanding the principles by
which each IS element selects its insertion target, it has already
become obvious that IS mediated mutations do not occur at random
along the genome.

Excision. It should be mentioned here that excision of some IS
elements has been reported. This process can be either precise or
nearly precise. It does not seem to depend on transpositional
functions nor in a strict way on recA dependent homologous
recombination (Foster et al. 1981). However, it could be that very
short segments of homology do play some role in excision. Such
segments are indeed often carried on both sides of an IS element
since replicombination has been shown for all studied IS elements to
lead to the duplication of the selected target over a length of 2-13
basepairs depending on the particular IS element (see Iida et al.
1983).

Consequences of DNA rearrangements. Excision of an IS element having
previously caused a DNA rearrangement of non-homologous genetic
material may lead to the fusion of two independent gene segments to a
single gene with novel functions or to the fusion of different genes
into a single operon securing the coordinate expression of a group of
newly linked genes. The postulated direct regulatory effects of IS
insertion are still somewhat controversial. However, it is now known
that some IS elements carry transcriptional promoters or terminators
which may have a direct influence on adjacent genes altering their
expression either positively or negatively. In other instances, it
has been shown that the insertion of an IS element did not bring
about directly the expression of adjacent genes. Rather by a polar
effect the insertion caused loss of functions of an adjacent gene.
But on studying revertants to activity, it was seen that in some
instances secondary DNA rearrangements, e.g. bringing about short
sequence duplications at the end of an IS element, had led to the

creation of a new promoter signal (Ghosal et al. 1978). Such small
DNA rearrangements may often be independent of transposition
functions.

Transposons. Transposons are defined as mobile genetic elements
containing one or several genes unrelated to transposition functions.
For convenience of easy selection, drug resistance transposons have
most thoroughly been investigated, but in principle any gene can be
found on a transposon. Some transposons are formed by two identical
IS elements flanking a segment of genetic information unrelated to
transposition. IS1 flanked transposons e.g. have been generated
stepwise in the laboratory by inserting subsequently two IS1 elements
on each side of a selectable gene. The resulting transposons carried
two IS1 elements either as tandem repeats or as inverted repeats and
both types of structure behaved as transposons (Iida et al. 1980).
Replicombination of such a transposon leads to a gene duplication. It
also offers the possibility to insert in principle any chromosomal
gene having become part of a transposon into a natural gene vector,
such as a viral DNA molecule or a transferable plasmid. These then
can spread the transposable gene in horizontal transmission to other
bacterial strains. If the infected recipient strain allows the
functioning of the IS element driving the transposon, incorporation
of the transferred gene into the host chromosome by transposition may
subsequently be expected.

Gene amplification. Genes carried by IS-flanked transposons have been
reported to occasionally duplicate and later to further amplify
(Meyer and Iida 1979). One possible mechanism to explain such
amplification would be the recombination between the right-sided IS
element of the transposon carried on one DNA molecule with the left-
sided IS element of the same transposon carried on a sister DNA
molecule. For obvious reasons, such gene amplification may offer
possibilities for divergent evolution.

Bacteriophage genomes as transposable genetic elements. Some
transposable genetic elements having the properties of a transposon
are endowed with all the functions necessary to be also a virus.
Bacteriophage Mu is the best studied example (Bukhari 1976). In its
vegetative growth transposition functions are expressed and cause a
burst of DNA rearrangements. As a consequence of its particular way
of replication, Mu carries in its phage particles segments of host
chromosomal DNA. Since Mu belongs to the temperate phages it can
lysogenize its bacterial host upon infection. It does so by
transposing into a variable target site of the host chromosome,
thereby often causing a mutation. This property brought this phage
its name Mu for mutator.

Evolutionary relevance of transpositional DNA rearrangements. In
summary, transpositional activities of IS elements and other movable
genetic elements can lead to a multitude of DNA rearrangements. With
a few exceptions, e.g. in the propagation of phage Mu, they occur
quite rarely and are hardly of any relevance for the average single
cell in a bacterial culture. Their relevance can be seen in the
evolutionary implications for populations. In pure populations of a
single microbial strain transpositional activities may influence to a

high degree vertical evolution. In mixed populations of
microorganisms and with the help of natural vectors, transposition is
obviously also of primary importance for horizontal gene transfer. In
both vertical and horizontal evolution subsequent events of DNA
rearrangement may give the genome the chance to try out various gene
combinations to form new operons. On the other hand, the fusion of
gene segments may in some cases lead to the formation of a new
enzymatic activity or of a multifunctional enzyme with activities
seen elsewhere on separate gene products.

The relevance of these conclusions, particularly with regard to
horizontal gene transfer, is testified by the results of a large
number of studies of the molecular basis of the mechanisms by which
microbial antibiotic resistance became a worldwide problem since the
intensive use of antibiotics during the last 30 - 40 years. Let us
recall that multiple drug resistance of many previously not resistant
bacterial strains began to be observed around 1960, about 10 years
after antibiotics had started to be widely used. Subsequent studies
then revealed first that often the genetic determinants for the
resistance were carried on natural gene vectors, in particular on
plasmids transferable in cell to cell contact. More recent studies
revealed in addition that quite frequently a single gene or a cluster
of drug resistance genes was carried on a transposon. This situation
illustrates convincingly that transposition indeed plays the
evolutionary role outlined in this chapter. Obviously, the drug
resistance genes must have been picked up in some living organisms,
were then transferred to other organisms via natural processes of
horizontal genetic exchange. This kind of horizontal gene flux
obviously had always occurred in nature. However, the worldwide use
of antibiotics in therapy and particularly in veterinary prophylaxis
drastically changed the selective condition in the natural
environment of many microorganisms. The result was a selective
enrichment for drug resistant bacteria. One can estimate that the
evolutionary adaptation of the microbial world to this newly created
environmental situation occurred in a timespan of about 10 - 20
years. This is a very informative result on how efficient horizontal
gene transmission can be, although the individual events of such
transmission are very rare compared to the lifespan of a single
bacterial cell in a propagating culture.

COMPARISON OF MECHANISMS CONTRIBUTING TO SPONTANEOUS MUTAGENESIS

Besides the transpositional processes discussed above, a variety of
other mechanisms are also relevant for spontaneous mutagenesis and
therefore for biological evolution. The information given in Table 1
on these mechanisms is probably incomplete due to lack of knowledge.
We particularly consider the randomness of events both with respect
to the site on the genome affected and with respect to the time of
the occurrence of the event. We also consider the relevance of a
mutational mechanism for vertical and for horizontal evolution.

Replication errors. One class of relatively well studied spontaneous
mutations is attributed to replication errors. These include
nucleotide substitution, deletion and insertion of one or a few

nucleotides as well as short tandem duplications. In many microbial strains, replication errors can be efficiently repaired so that only errors escaping this repair are relevant for spontaneous mutagenesis. It is generally assumed that replication errors occur at random, at least on the large scale of a bacterial genome, and also at random in time.

Mutagen-induced alterations. Defects molecularly similar to replication errors can be brought about by classical mutagens, e.g. some chemicals. These probably should not be considered to occur fully at random in time. Their action along the bacterial genome may often be random, although some mutagens may preferentially act at A-T rich regions or at single stranded DNA segments.

The defects affecting one or a few nucleotides listed under A and B in Table 1 have their evolutionary relevance in the vertical gene transmission, but they can generally not be considered as of direct relevance in horizontal gene transfer. Exceptions to this rule exist, e.g. if a viral vector particle is exposed to an external mutagen, or if a nucleotide sequence alteration fortuitously results in a crossover site for a site specific recombination system belonging to a natural gene vector.

Mutations resulting from recombination. Several recombinational processes may also cause spontaneous mutagenesis. Many bacteria are proficient in general recombination, although their propagation may not be directly dependent on this activity, except for purposes of repair. However, upon horizontal gene transfer, e.g. in conjugation, transduction, transformation and cell fusion, two more or less closely related bacterial strains may exchange part of their chromosomes. Reciprocal recombination at structurally and functionally homologous regions of the genomes may then lead to the formation of hybrids (class C of Table 1). Such processes might often not be expected to lead to viable hybrids if the participating bacteria are only distantly related, or such hybrids may show a serious growth disadvantage relative to their parents. In these instances they will escape easy detection.

The events of reciprocal recombination alluded to in section D of Table 1 are not meant to affect the bacterial genome as a whole. The processes enlisted here could be seen to bring about either deletion formation or gene duplication, if the reciprocal recombination occurred intramolecularly or between two sister-chromosomes. Alternatively, the same processes may be considered to occur between two independent replicons and to lead to cointegration, e.g. between the bacterial chromosome and a plasmid or a viral genome. The role of an IS element in these processes is completely independent of its transpositional activities. Rather the IS element should be considered here as playing a passive role in just providing a portable segment of homology to two otherwise unrelated segments of genetic information. Because of the considerable length of IS elements this mechanism can occur with frequencies ranging up to in the order of 1% per cell generation. Such recombination can upon subsequent excision or other deletional loss of the IS element lead to the direct fusion of unrelated genes or segments of genes.

Several reports testify of the rare use of short segments of chance
homology down to about ten base pairs in length in reciprocal
recombination (Farabaugh et al. 1978). It seems that these
recombinations are sometimes brought about by mechanisms requiring
the recA product. As already mentioned above in the section on
excision of an IS element, it seems that some recA independent
processes may lead to the same result.

An increasing number of investigations have revealed the existence of
various site specific recombination systems. The classical example
for this class is the integration of the λ phage genome into the
bacterial chromosome during lysogenisation (Nash 1981).

Recombination at non-homologous sequences (class E, Table 1) is very
rare. DNA gyrase has been reported to have affinity for some specific
sites on the DNA and to depend in its function on more than one
subunit. It has been postulated that the interaction of two subunits
each fixed at different sites on DNA may lead to the recombination of
non-homologous sequences (Ikeda et al. 1982). Other recombinations at
non-homologous sequences are perhaps "illegitimate" events to be
defined as erroneous and irreproducible. At the time being we also
include into the class of illegitimate recombinations any event for
which we have no likely mechanistic explanation yet. Note, however,
that the term "illegitimate" is used by some authors for all genetic
recombinations not based on processes of general recombination
occurring between largely homologous DNA molecules.

Transpositional replicombination (class F, Table 1) has already been
discussed in detail above. In contrast to most of the other
mechanisms contributing to spontaneous mutagenesis, we did not list
transposition as occurring random in time. However, this is not based
on any strong evidence since no means for massive induction of
transposition has yet been found. But it might be that a variety of
both internal and external factors may affect the occurrence of
transpositional processes, and such inducing factors might not be
constant as a function of time. We also did not list transposition to
occur at random locations on the genome, since many transposable
elements seem to have their own, specific criteria for target
recognition as already outlined. A few of the other recombinational
mechanisms are likely also not to occur at random along a genome,
while others may perhaps do so.

Most of the recombinational mechanisms for spontaneous mutagenesis
can be considered to act both in vertical and horizontal gene
transmission as indicated in Table 1.

Relative frequencies and consequences of various mutagenic processes.
An important question has not been alluded to yet. This is to know
the relative importance of each of the different mechanisms causing
spontaneous mutagenesis. This question is difficult to answer since
direct comparative measurements are difficult and often impossible to
make due to technical reasons. From studies made with E. coli
bacteria carrying the genome of bacteriophage P1 as a plasmid, it can
be concluded that transposition is the most frequent cause of
spontaneous mutagenesis of the phage P1 genome as a whole (Sengstag

TABLE 1. Mechanisms contributing to spontaneous mutagenesis in bacteria

Mechanism — Class / Specific example	Randomness with regard to — Genome location	Randomness with regard to — Time	Relevance for mutation in — Vertical gene transmission	Relevance for mutation in — Horizontal gene transmission

A. Replication error:
  – Nucleotide substitution
  – Deletion of one/few nucleotides
  – Insertion of one/few nucleotides
  – Short tandem duplication
	+	+	+	–

B. Effect of external mutagen (chemicals etc.):
  – Nucleotide substitution
  – Small deletion / insertion / duplication
  – Other alterations
	+/–	+/–	+	+/–

C. Reciprocal recombination between chromosomes of more or less closely related species	+/–	+/–	–	+

D. Reciprocal recombination between segments of homology in largely non-homologous DNA:
  – Homology provided by an IS element (~ 1000 bp)
  – Chance homology extending over short sequence
  – Site specific recombination system
	–
	+/–
	–
		+/–
		+/–
		+/–
			+
			+
			+
				+
				+
				+

E. Recombination at non-homologous sequences:
  – Mediated by DNA gyrase
  – Other, "illegitimate"
	+/–
	+/–
		+
		+/–
			+
			+
				+
				+

F. Transpositional replicombination, resulting in:
  – Transposition of IS element
  – Cointegration
  – Deletion
  – Inversion
	–	–	+	+

and Arber 1983). Most of the P1 mutants thereby considered carry
lethal mutations, but can be propagated and studied while the genome
is kept in the prophage state. One may expect that alterations
affecting one or only a few nucleotides are less often lethal than
larger DNA rearrangements. Such alterations may sometimes not result
in any phenotypic change, but be of evolutionary relevance, of
course. Therefore, for small, stepwise alterations of the genetic
message these "point mutations" are probably quite relevant. In
contrast, recombinational alterations can be considered as bringing
about large evolutionary steps by single events. This is of
particular relevance in horizontal gene flux allowing a recipient
cell to acquire entire functional genes or functional parts of a gene
from other living organisms. In vertical as well as in horizontal
evolution DNA rearrangements can bring about the genesis of new
functional units, genes and operons, as already outlined.

In considering the contributions to evolution to be expected by the
different recombinational mechanisms, one has to take into account
the dependence of each mechanism on particular nucleotide sequences,
representing either homology, crossover sites for site specific
recombinations or target sites for transposition. These requirements
represent limitations to the number of possibilities for new DNA
rearrangements and this may result in some degree of reproducibility
of these natural, evolutionary processes. Less frequently occurring
recombinations between non-homologous sequences, however, leave some
possibilities to occasionally overcome such limitations inherent to
some of the major recombinational mechanisms. This offers
opportunities for rare kinds of DNA rearrangements not inherent to
other more often used mechanisms such as general or site specific
recombination.

It should be recalled here that bacteria as other living organisms
possess a number of strategies for species isolation. These represent
natural borders to horizontal gene flux. Among these strategies are
the barrier formed by cellular membranes to DNA penetration, host-
specific DNA restriction and modification systems and cell dependence
of the mechanisms guiding gene expression. As is well known, such
limitations also show some leakiness which allows for occasional
violation of existing general rules. This complex situation can be
considered as an optimal equilibrium between a strict genetic
isolation of particular strains of microbes and a general gene pool
which would allow for frequent genetic exchange at the molecular
level.

          COMMON FEATURES OF RECOMBINANTS PRODUCED EITHER IN VIVO OR
          IN VITRO

The various natural recombinational mechanisms are important
components of horizontal evolution, both for the loading of genetic
information onto natural gene vectors and for the incorporation of
transferred genes into the recipient chromosome. IS elements thereby
play a double role: a) actively by their transpositional activities,
and b) passively by providing portable regions of homology.
Recombinants which are formed by either of these two routes will
still carry the IS element between the two segments of usually

largely non-homologous DNA sequences. Subsequently occurring independent events such as excision of the IS element or also a rare illegitimate recombination may lead to the loss of all or part of the IS element. This allows the fusion of genetic material not representing IS sequences. If such secondary DNA rearrangements occur to a transposon while it is still carried on a natural vector, the resulting structure may render the integration of a transferred gene into a new recipient host chromosome somewhat less likely, except if the chromosome carries homology with the newly acquired gene. The former type of structure can be compared with in vitro recombinant DNA molecules representing hybrids between a vector DNA and a chosen segment of genetic information. This segment, of course, will usually not contain an IS element to provide transpositional mobility. In this sense, recombinant DNA molecules are less apt than many of the encountered wild type drug resistance plasmids to donate their carried genes to infected recipients for stable integration. But, as we have seen, cointegration mediated by a chromosomal IS element and other mechanisms of recombination between largely non-homologous sequences may bring about occasional hybridisations between recipient genomes and non-transposable genes which may have been introduced into a cell by appropriate vector molecules. Also, as we have already outlined, any genome segment may become part of a transposon being generated by transposition of IS elements. These mechanisms might of course also affect cloned foreign genes propagated in host bacteria.

Another similarity between in vitro recombinant DNA molecules and in vivo DNA rearrangements due to mobile genetic elements or other recombination processes can be seen in the following. Not all new recombinants fit well into an existing physiological harmony of the host cell. Some of the recombinations may inhibit the cell to propagate further or they may at least cause a growth disadvantage. Particularly the latter situation is a typical condition, in which secondary spontaneous mutations may later restore normal growth, allowing for selective enrichment of mutated subclones. Any of the mechanisms outlined in Table 1 may contribute to such evolutionary gene modulations, which may in fact often not be noticed, except by attentive investigators or upon nucleotide sequence verification.

Our considerations of the importance of horizontal gene transmission for biological evolution in the microbial world have two consequences with regard to the question of biosafety of recombinant DNA technology. On the one hand the identification of various mechanisms involved in the promotion of horizontal gene transmission clearly shows that this principle is widespread in nature. It may perhaps even allow genetic exchange to occur between prokaryotes and eukaryotes. One can imagine that in sequential steps of horizontal gene transmission relatively large distances of genetic relatedness can be bridged. For these reasons, the risk to produce absolutely novel types of recombinants by in vitro recombinant DNA technology becomes less likely. On the other hand, however, we have to be aware that any gene transferred in vitro into E. coli will not necessarily stay only in this particular bacterial strain. Rather the mechanisms of horizontal gene transmission may mobilize such gene and eventually spread it to a number of different microorganisms if the opportunity

for growth in mixed populations is given. This knowledge is relevant
for any question regarding biohazards related to deliberate or
accidental release of recombinant DNA molecules into the environment.

RELEVANCE OF THE TOTAL GENE POOL FOR BIOLOGICAL EVOLUTION

A general conclusion to be reached from the knowledge of horizontal
gene transmission is that the future biological evolution must be
considered not to solely depend on subsequent mutations occurring in
each particular branch of the evolutionary tree during vertical gene
transmission. Rather horizontal gene transmission may lead to major
steps in the evolution of any branch of the tree. Therefore the
total gene pool of all microorganisms taking occasionally part in the
horizontal exchange of genetic information is important for future
evolution of each of today's species. One might speculate that this
conclusion may also be relevant for higher organisms.

ACKNOWLEDGMENTS

The author thanks Shigeru Iida and Brian Dalrymple for stimulating
discussions during the preparation of the manuscript and the Swiss
National Science Foundation for continued support.

REFERENCES

Bukhari, A.I. (1976). Bacteriophage Mu as a transposition element.
    Annu. Rev. Genet. 10, 389-412.

Campbell, A., Berg, D.E., Botstein, D., Lederberg, E.M., Novick,
    R.P., Starlinger, P. and Szybalski, W. (1979). Nomenclature
    of transposable elements in prokaryotes. Plasmid 2, 466-
    473, and Gene 5, 197-206.

Engler, J.A. and van Bree, M.P. (1981). The nucleotide sequence and
    protein-coding capability of the transposable element IS5.
    Gene 14, 155-163.

Farabaugh, P.J., Schmeissner, U., Hofer, M. and Miller J.H. (1978).
    Genetic studies of the lac repressor. VII. On the molecular
    nature of spontaneous hotspots in the lacI gene of
    Escherichia coli. J. Mol. Biol. 126, 847-863.

Foster, T.J., Lundblad, V., Hanley-Way, S., Halling, S.M. and
    Kleckner, N. (1981). Three Tn10-associated excision events:
    Relationship to transposition and role of direct and
    inverted repeats. Cell 23, 215-227.

Galas, D.J., Calos, M.P. and Miller, J.H. (1980). Sequence analysis
    of Tn9 insertions in the lacZ gene. J. Mol. Biol. 144, 19-
    41.

Galas, D.J. and Chandler, M. (1981). On the molecular mechanisms of
    transposition. Proc. Natl. Acad. Sci. U.S.A. 78, 4858-4862.

Ghosal, D., Gross, J. and Saedler, H. (1978). DNA sequence of IS2-7 and generation of mini-insertions by replication of IS2 sequences. Cold Spring Harbor Symp. Quant. Biol. 43, 1193-1196.

Iida, S., Meyer, J. and Arber, W. (1980). Genesis and natural history of IS-mediated transposons. Cold Spring Harbor Symp. Quant. Biol. 45, 27-43.

Iida, S., Meyer, J. and Arber, W. (1983). Prokaryotic IS elements, in Mobile genetic elements (Shapiro, J.A. ed.) pp. 159-221, Academic Press, New York.

Ikeda, H., Aoki, K. and Naito, A. (1982). Illegitimate recombination mediated in vitro by DNA gyrase of Escherichia coli: Structure of recombinant DNA molecules. Proc. Natl. Acad. Sci. U.S.A. 79, 3724-3728

Meyer, J. and Iida, S. (1979). Amplification of chloramphenicol resistance transposons carried by phage P1Cm in Escherichia coli. Mol. Gen. Genet. 176, 209-219.

Meyer, J., Iida, S. and Arber, W. (1980). Does the insertion element IS1 transpose preferentially into A+T-rich DNA segments? Mol. Gen. Genet. 178, 471-473.

Nash, H.A. (1981). Integration and excision of bacteriophage λ: The mechanism of conservative site specific recombination. Annu. Rev. Genet. 15, 143-167.

Sengstag, C. and Arber, W. (1983). IS2 insertion is a major cause of spontaneous mutagenesis of the bacteriophage P1: non-random distribution of target sites. EMBO J. 2, 67-71.

Shapiro, J.A. ed. (1983). Mobile genetic elements. Academic Press, New York.

Starlinger, P. and Saedler, H. (1976). IS elements in microorganisms. Curr. Topics Microbiol. Immunol. 75, 111-153.

DISCUSSION

W. ZILLIG: are there any good examples for horizontal gene exchange over a larger phylogenetic distance, aside from special cases like Agrobacterium in plants?

W. ARBER: The best example I know is the observed wide spreading of antibiotic resistance genes.   Unfortunately, the original hosts of these genes remain largely unknown, so that we do not have good data on the horizontal distances over which these genes were transferred.

P. STARLINGER: Why did ampicillin-resistant Neisseria arise only a couple of years ago, while gonorrhea has been treated with this drug since World War II?

W. ARBER: I think Stuart Levy is more competent to answer this.

S.B. LEVY: It is certainly puzzling that it took this long, but it may take a significant evolutionary time for this rare genetic exchange to occur.   Penicillin use has dramatically increased since the 1940's. Moreover, the origin of gonorrhea resistance to ampicillin came from areas (brothels) where the drug was being used prophylactically.

TRANSITORY RECOMBINATION BETWEEN PHAGE AND PLASMID GENOMES.

S.D. Ehrlich, M. Dagert, S. Romac and B. Michel

Institut Jacques Monod, C.N.R.S., Université Paris VII,
2, Place Jussieu, 75251 Paris cedex 05, France.

INTRODUCTION

Homologous DNA recombination re-assembles genes, whereas illegitimate
recombination changes the gene order. This latter process thus
creates genetic diversity, which is of importance in the evolution
of living organisms. On a shorter time-scale, the process modifies
the structure of many genomes constructed in vitro, and is therefore
of great relevance in gene cloning. It can also change gene
expression, which in the case of oncogenes, is of medical importance.
In spite of its interest not much is known about illegitimate
recombination, besides what has been learnt about specialized
genetic elements such as transposons and insertion sequences
(Kleckner, 1981).

We have studied recombination between the phage M13 and the plasmid
pHV33, which have no homology longer than 13 bp. The two genomes
combine in *Escherichia coli* and form a chimera which is encapsidated
in phage proteins. When the chimera is introduced into a recipient
cell the two parental genomes are regenerated. We call the
regeneration process "decombination" and the entire combination-
decombination process "transitory recombination", to denote the
fact that the chimeras do not exist "permanently" but only
that host functions for homologous recombination are also involved
in this process. We speculate that it may be one of the first phases
of both homologous and illegitimate recombination.

Recombination between f1, a phage similar to M13 and another plasmid,
pSC101, which has been studied previously, did not involve transitory
recombination, but rather co-integrate formation during transposition
of the plasmid-carried insertion sequence IS101 onto the phage
(Ohsumi et al., 1978, Ravetch et al.,1979; Fischoff et al., 1980).

RESULTS

Nonhomologous phage and plasmid genomes undergo transitory
recombination.
Plasmid pHV33 (Primrose and Ehrlich, 1981), a hybrid between pBR322
and pC194 (Iordanescu, 1975), and phage M13 are maintained in
E. coli cells as independent, compatible, replicons. Phage particles
excreted from cells harboring the two genomes contain plasmid

sequences at a frequency of 10^{-4} relative to phage sequences, as judged both by biological (transduction for plasmid-carried genetic markers) and biochemical (hybridization) assays. The following evidence indicates that, in a transducing particle, plasmid and phage genomes are covalently linked.
(1) Plasmid sequences are encapsidated in M13 proteins, since the transducing activity and infectivity of the phage stock are inactivated in parallel by M13 antiserum.
(2) Plasmid sequences are single-stranded in the transducing particles, since in the CsCl gradients they band at the same density as phage particles.
(3) Transducing particles are twice as big as phage particles, as shown by sucrose gradient velocity sedimentation. This is as expected if they contain one phage and one plasmid genome, since pHV33 (7.2 kb) is not much bigger than M13 (6.4 kb; Wesenbeek et al., 1980).
(4) A molecule bigger than the M13 genomes contains plasmid sequences as shown by electrophoretic analysis of DNA isolated from the transducing phage stock. The mobility of plasmid DNA was determined either by transformation (Harris-Warwick et al., 1975) or hybridization (Southern, 1975) tests, that of phage DNA by its fluorescence or its transfecting activity.
(5) Plasmid and phage genomes are not joined by a protein, since the treatment with SDS, phenol and proteinase K of the DNA extracted from a transducing phage stock did not modify the phage-plasmid linkage in a transformation assay.
(6) Plasmid and phage are covalently linked, since (a) annealing of the single-stranded DNA extracted from the transducing stock with the M13 complementary strand DNA converts molecules containing plasmid sequences to a (partially) double-stranded form, as judged by chromatography on hydroxyapatite (Bernardi, 1969); (b) cleavage within the double-stranded portion of such molecules by a restriction endonuclease does not release single-stranded plasmid sequences, as judged by hydroxyapatite chromatography. Phage-plasmid linkage is therefore not topological (the two genomes are not simply catenated) but covalent.

Cells infected by transducing particles, or transformed by DNA extracted from such particles, do not contain chimeric genomes. They always harbor a plasmid indistinguishable from the authentic pHV33 by genetic (presence of all plasmid-specified markers in over 10^3 tested transformants) and biochemical criteria (size, restriction analysis, which would detect any difference exceeding 30 bp in 20 analysed DNAs). In 70% of the cases they also harbor a phage indistinguishable from M13 by plaque morphology and restriction analysis (in 18 out of 20 cases; the remaining two phages carried a small modification in the intergenic space).

The data presented show that two non-homologous genomes combine with a rather high efficiency (10^{-4}, compared with 10^{-5} or less observed with most transposons; Kleckner, 1981), exist linked in a chimera, and then decombine. This last phase occurs with a probability approaching unity, as judged from the fact that the ratio of encapsidated plasmid to phage sequences and the ratio of transductants to viable phage are both close to 10^{-4}. It probably occurs rapidly, since the chimeras are not viable in *E. coli* (as judged from experiments with hybrids constructed *in vitro* between M13 and pHV33) and, in order to give a transductant, have to decombine before being lost from the transduced cells. The combination-decombination process of pHV33 and M13 we call transitory recombination.

Genetics of transitory recombination. We tested the effect of different known recombination mutants first on the combination, then on the decombination phase of transitory recombination. In the first series of experiments phage stocks were prepared by infecting different *E. coli* mutants harboring pHV33 and used to transduce the plasmid into the wild-type recipient cells. In the second series a phage stock grown on the wild-type cells harboring pHV33 was used to transduce the plasmid into different mutant cells. The control experiments showed that (i) the copy numbers of phage and plasmid in different mutant cells did not vary greatly; (ii) different mutants supported growth of M13 and replication of pHV33 with similar efficiency; (iii) all mutants were infected with similar efficiency by M13 phage; (iv) pHV33 could be established in different mutants with equal efficiency.

The results show that actitivy of either *recBC* or *recA* together with *recF* (and possibly *recL*) genes is necessary for the formation of chimeras. This indicates the existence of two combination pathways (*recBC* and *recAF*). The *recAF* pathway is less active in strains carrying the *sbcA* mutation.

Decombination of chimeras is inhibited by the *recBC* mutation. The inhibition is not suppressed by the known suppressors of *recBC*, *sbcA* or *sbcB*, but is suppressed by the *recL* mutation. These results indicate the existence of two decombination pathways, the *recBC* pathway and a pathway blocked by the *recL* gene product. Chimeras formed by either the *recBC* or *recAF* pathway can decombine along both the *recBC* and the alternative pathway.

Certain genes influencing homologous recombination can also influence transitory recombination. This suggests that the two processes have some steps in common.

OPEN QUESTIONS

The concept of transitory recombination raises many questions :
(1) is the process general or restricted to the two genomes studied?
(2) can any sequence engage in transitory recombination?
(3) what is the signal for decombination?
(4) what is the relation between transitory and homologous recombination? Between transitory and illegitimate recombination?
(5) what is the biological significance of transitory recombination?

S.D. EHRLICH et al.

Although they remain to be proven, our current working hypothese are:
Transitory recombination is the first phase of homologous recombination. Molecules, combined in a chimera, explore each other for
homology. If homology is found they engage in homologous recombination, if not, they decombine. Illegitimate recombination occurs if
the decombination mechanism fails.

ACKNOWLEDGMENTS

S.D.E. is on the INSERM research staff, M.D. was the recipient of
a fellowship from Universidad de Los Andes, Merida, Venezuela,
S.R. was a French-Yugoslav exchange fellow. The work was supported,
in part, by INSERM grant n° 821009.

REFERENCES

Bernardi, G. (1969) Biochim. Biophys. Acta. 174, 423-434.
Harris-Warwick, R.M., Elkana, Y., Ehrlich, S.D. and Lederberg, J.
        (1975) Proc. Natl. Acad. Sci. U.S.A. 72, 2207-2211.
Iordanescu, S. (1975) J. Bacteriol. 124, 597-601.
Kleckner, N. (1981) Ann. Rev. Genet. 15, 341-404.
Oshumi, M., Vovis, G.F. and Zinder, N.D. (1978) Virology 89,438-449?
Primrose, S.B. and Ehrlich, S.D. (1981) Plasmid 6, 193-201.
Ravetch, J.V., Oshumi, M., Model, P., Vovis, G.F., Fischoff, D. and
        Zinder, N.D. (1979) Proc. Natl. Acad. Sci. U.S.A.
        76, 2195-2198.
Wesenbeek, P.M.G., Hulsebas, T.J.M. and Schoenmakers, J.G.G. (1980)
        Gene 11, 129-148.

Fischoff, A.D., Vovis, G.F. and Zinder, N.D. (1980) J. Mol. Biol.
        144, 247-265.
Southern, E.M. (1975) J. Mol. Biol. 98, 503-517.

DISCUSSION

P. STARLINGER: Do you envisage an initial contact of DNA molecules
that does not use homology even in homologous recombination?

S.D. EHRLICH: Yes.   The current model is that the initial contact
between the recombining genomes occurs irrespective of homology, and
that the homology search then follows.

W. ARBER: To what extent may the phenomenon which you described depend
specifically on the biology of bacteriophage M13?

S.D. EHRLICH: We cannot rule out that possibility.   Plasmid
transduction reported in other E. coli, as well as S. aureus and B.
subtilis systems resembles, however, the phenomenon we described.   In
all cases a non-modified plasmid was transduced.   We are actively
engaged in searching for evidence that transitory recombination is a
general phenomenon.

SURVIVAL OF PLASMIDS IN ESCHERICHIA COLI

S.B. Levy
Departments of Medicine and of Molecular Biology and Microbiology, Tufts University Medical School, Boston, Mass 02111 U.S.A.

Plasmids in E. coli, especially those of the non-conjugating and/or poorly-mobilizable types, are commonly used vectors for gene cloning. The survival of these plasmids can be assessed by the survival of their host E. coli and by their spread to other bacteria in the environment.

E. coli is normally found in the large intestine of man and other warm-blooded animals. Certain strains occasionally find their way into the urinary tract and other body tissues. They may be the cause of bacteremia and sepsis, particularly in immunocompromised individuals. They have been isolated in soils, in water (especially where fecal contamination is evident), from insects including the common fly, as well as in birds and fish. While E. coli are natural inhabitants of animal intestines, a particular E. coli strain will not colonize all species. However, data also demonstrate that there is no host specificity. Colonization depends on many conditions, including resident bowel flora and physiologic state of the infecting E. coli (1). Wild-type and laboratory E. coli, except the bile acid sensitive derivatives such as χ1776 (2), will always colonize germ-free animals. In addition to competition with bowel flora, other critical determinants of stable maintenance in the intestine include motility of the infecting strain, the host's intestinal peristaltic activity, and secretions of the bowel.

Survival of wild-type E. coli and debilitated E. coli K12 in the intestinal tract of man and other mammals. Initial studies in our laboratory examined the survival of severely-disabled E. coli K12 strain χ1776 as compared to a normal E. coli K12 in germ-free mice and in human subjects. There was no detectable survival of the debilitated strain in either mammal 24 hours after ingestion. However, following ingestion of 2 x 10^{10} organisms in milk, human volunteers did excrete strain χ1776 bearing plasmid pBR322 for 4 days in an amount equal to 6 organisms for every million ingested (Table). The normal E. coli K12 χ1666 (ara⁻) easily colonized the mice but not man (3). This strain, with or without the plasmid

RECOVERY OF E. COLI HOST-VECTOR SYSTEMS IN HUMAN INGESTORS

Test Organism	Plasmid	Dose Ingested	Range of Recovery Rates	Mean Recovery Rate $\pm$ SD
χ1776		3.8×10^9 1.5×10^{10}	$5 \times 10^{-6*}$ to $<10^{-8}$ $<10^9$	--
χ2236 (χ1776)[a]	pBR322	2.3×10^{10}	$9.1-1.3 \times 10^{-6}$	$5.9 \pm 3.87 \times 10^{-6}$
χ1666		8.4×10^9	2.0×10^{-2} to 1.6×10^{-4}	$0.75 \pm 1.1 \times 10^{-2}$
D20-5 (χ1666)	pBR322	9.4×10^9	2.0×10^{-1} to 8.7×10^{-3}	$5.9 \pm \times 10^{-2}$
D300-1 (χ1666)	pBR322 pLM-2 pSL222-4	5.0×10^9	6.4×10^{-1} to 9.6×10^{-5}	$1.6 \pm 3.2 \times 10^{-1}$

[a] parent strain in parentheses

* One five-hour excretion of 1.9×10^4 in one ingestor only

pBR322, was recovered in ten thousand-fold greater numbers in man and for two days longer than was χ 1776 bearing pBR322. There was somewhat better recovery of the plasmid-containing derivative (3). The surviving bacteria during this 3-4 day period represented .01% of the total E. coli in the feces. The numbers did not substantiate intestinal colonization, and excretion did not last more than 4 days. The plasmid survived no better. Using the antibiotic resistance markers present on pBR322, and DNA:DNA hybridization to the tetracycline resistance determinant carried by pBR322, we were unable to demonstrate any transfer of this plasmid to indigenous Gram-positive or Gram-negative aerobic flora of the volunteers (3). Others have also shown lack of colonization of E. coli K12 in man, and poor, if any, transfer of its conjugatable plasmids to endogenous gut flora (4,5).

Plasmid transfer to human strains of E. coli was also examined in the intestine of germ-free mice. After the mice were colonized with 10^{10} cfu/gm of D20-5, (χ 1666 containing pBR322), they received 2 x 10^8 viable cells of 4 different wild-type human E. coli mixed in equal proportions. These were able to ferment arabinose and so could be distinguished from D20-5. Three strains showed no antibiotic resistance; the fourth was resistant to streptomycin and chloramphenicol. The E. coli population established at a level of 10^{11} cfu/gm in about equal amounts of all four strains. In this specially derived densely-populated mammalian system, we found no transfer of pBR322 to any of the four human strains.

In order to increase the possibility of transfer of pBR322 to indigenous flora, we introduced two conjugative mobilizing plasmids of different incompatability types, plasmid pLM2 (inc P) and pSL222-4 (inc FII) into D20-5 to derive strain D300-1. All plasmids had detectably different antibiotic resistance markers. Germ-free mice were colonized initially with human feces and then tested to determine if any organisms were resistant to the combination of tetracycline (tet) and ampicillin (amp) (as on pBR322), or to kanamycin (kan) alone (as on pLM2), or to chloramphenicol (cam) (on pSL222-4), or to a combination of tet, amp, kan and cam. None of the bacteria colonizing these mice showed resistance to these antibiotics. Into this intestinal flora system, 5×10^9 viable D300-1 were introduced, and daily samples of stools were examined for the detection of transfer of pBR322 from D300-1 to the human E. coli. D300-1 reached stable titers of 10^2-10^3 cfu/gm while the human flora remained stably at 10^7-10^8 cfu/gm. There was no evidence of plasmid transfer to indigenous bacteria. Furthermore, no transfer of pBR322 was noted, even after the animals were given 400 mg/l of tetracycline and again associated with D300-1 for a period of 5 days.

Studies of D300-1 were next carried out in humans in order to determine if pBR322 could be transferred from D300-1 to gut flora. D300-1 was recovered at rates of 10^{-1} to 10^{-4} over 3.5 to 6 days (6) (Table). Despite the administration of tetracycline to the volunteers, this K12 strain did not increase its titer and thus appeared to be at a selective disadvantage in the human gut even though it carried a tetracycline resistance plasmid. Transfer of pBR322 was not detected, but transfer of another resident plasmid, the transfer-derepressed plasmid, pSL222-4 (Fig. 1) was observed. Even in this case, the latter plasmid transferred 10^{-4}-fold less frequently than it did in vitro. Although D300-1 was unaffected by tetracycline ingestion, the number of pSL222-4 transcipients was enhanced (Figure 1). Despite efficient transfer in plate matings, the inc P plasmid pLM2 was not detected in any indigenous host bacteria. Using these data and the estimate of the total number of bacteria colonizing the gastrointestinal tract, we estimated that any mobilization in vivo of pBR322 in E. coli K12 to an indigenous recipient could not occur at a frequency higher than 10^{-12}.

Levine and co-workers have examined the survival and transfer of plasmids in good colonizing human strains of E. coli (7). They took the poorly-mobilizable plasmid pBR325 and placed it in a human E. coli strain HS-4. When 5×10^{10} organisms were ingested by 15 volunteers, good colonization levels were achieved, but in no instance was pBR325 found in any indigenous bacterial flora. When pBR325 was placed in the same HS-4 host bearing a mobilizing derepressed plasmid (F-amp), it still showed no transfer to indigenous flora. Only when tetracycline was given to the volunteers taking the combined organism HS-4 (pBR325, F-amp), was transfer of pBR325 to gut flora demonstrated.

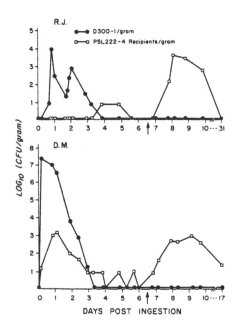

Figure 1. Recovery of indigenous E. coli containing pSL222-4. Daily fecal samples were examined for the presence of cam-resistant bacteria. These were subsequently isolated and checked for the presence of plasmid pSL222-4. Recoveries of less than 10 recipients/g of feces were made in broth cultures. Arrow denotes the initiation of tetracycline therapy (1.5 g orally over 1 1/2 days). (From ref. 6).

In order for a poorly-mobilizable or even mobilizable plasmid to be transferred out of its ingested host E. coli into other hosts, it must be mobilized by another plasmid. Thus, its host must pick up mobilizing plasmids during its transit in the gut. In our studies with E. coli K12, we detected no transfer of the poorly-mobilizable pBR322 to indigenous flora despite survival of the host bacteria in the gastrointestinal tract for 4-6 days at reasonably high titers (3). Using a highly colonizable strain (HS-4), Levine et al reported that this kind of phenomenon occurred with a mobilizable plasmid, but did not occur with a poorly-mobilizable plasmid (7).

These studies indicate that survival of plasmids in the human intestinal tract will be increased if they are in a wild-type colonizing E. coli, and additionally enhanced if the plasmid bears resistance to a drug which the individual is ingesting. In the absence of selective pressure or of ingesting large quantities of these organisms in a buffer, colonization and transfer are undetected. These conditions are highly unlikely to occur naturally, thus further limiting the possibility of such a rare genetic exchange.

This selective effect of antibiotics on emergence of plasmid recipients parallels results we found when studying wild-type E. coli in chickens. Biochemically-marked plasmid and E. coli were inoculated into four chickens, 2 of each were placed into different cages, with or without tetracycline-laced feed. In the tetracycline-fed cage, the plasmid and E. coli strain spread to the other chickens; in the cages given normal feed, the resistant strains were soon undetected (8).

Transfer of ` R factors and other plasmids in the mammalian gut is probably a common event (9,14-16). Still the frequency of transfer is low and the persistence of new recipients in the absence of selection is variable.

Factors Affecting Colonization of the Intestinal Tract by E. coli. There has been considerable effort extended towards understanding colonization by enteropathogenic E. coli in the diarrheal diseases. Outer membrane antigens appear to be involved in the attachment between bacteria and the epithelial cells lining the small intestine. The colonization of the large bowel by E. coli has not been extensively studied. Wild-type E. coli do not cross-colonize different individuals easily and even an isolate from the same individual may not stably recolonize when given back in an oral preparation (9). The factors which lead to "transient" and "resident" strains of E. coli in the gut (10) are still not understood. We performed some studies to determine whether there were particular sites of colonization for different E. coli in the gastrointestinal tract (11). E. coli K12 χ1666 (nalidixic acid resistant) colonizes germ-free mice to high levels. From this strain we made two congenic derivatives, one rifamycin resistant and the other, streptomycin resistant, both obtained by spontaneous mutation on antibiotic-containing plates. Upon separate feeding of these organisms to germ-free mice, all reached similar titers of colonization in the range of 10^8 cfu/gm. When we examined these three strains in the same mouse intestine, we found competition between the strains. Both mutant strains competed poorly with the parental strain and with each other (Figure 2). The same strain containing plasmid pBR322 was also a poor competitor. Thus, it would appear that bacterial strains show reduced ability to colonize in competition with congenic strains. We could correlate these differences with changes in proteins in the cell envelope (11).

In studies using more conventional mice, but streptomycin-treated, Cohen et al asked similar questions (12). These investigators were interested in knowing what factors were involved in E. coli colonization of the large bowel. They had demonstrated that by streptomycin treatment of conventional mice, they could eliminate the facultative anaerobes while maintaining obligate anaerobes, and thereby show improved colonization of these mice with particular strains of wild-type human E. coli. Using this system, they examined the relative colonizing ability of different human E. coli strains. Most of the wild-type E. coli colonized well, but some did not. They then examined the ability of two good colonizing strains to coexist in the mouse gut. In most cases they found that both organisms fed together would colonize the mouse gut in equal numbers. When, however, they derived rifamycin resistant mutants of these wild-types, and fed them back with their parental hosts, they observed a decreased competition as we had found with derivatives of

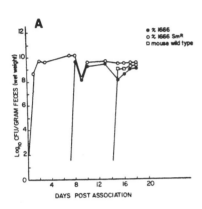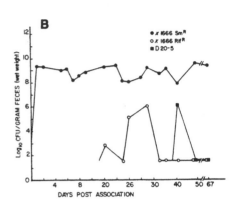

Figure 2. Colonization of germfree mice by (A) χ1666 Sm^r followed by χ1666 and a wild-type mouse E. coli; (B) χ1666 Sm^r followed by χ1666 rif^r and then D20-5 (from ref. 11).

E. coli K12 in germ-free mice. These mutant derivatives, however, which competed poorly with the parental types, had no problem competing with other human strains colonizing these streptomycin-treated mice. These studies with E. coli suggest that there are sites in the gut which may be strain-specific and which would, therefore, limit the kinds of E. coli capable of colonizing. The basis of this distinction has not been determined.

Notwithstanding the above, it is obvious that very little is known about the factors involved in E. coli colonization of the large intestine. Since this is the natural habitat of this organism, and since E. coli is a predominant host bacterium to be used in genetic engineering, studies along these lines are of critical importance in understanding and evaluating the capabilities of E. coli to survive in different animal species including man.

Survival of plasmids in E. coli in sewage waste disposal systems. Several groups have examined the factors affecting survival of E. coli and other organisms in the environment and, more recently, the effect of normal sewage disposal on the survival of plasmid vectors contemplated for use in recombinant DNA engineering experiments. It is well documented that clay serves as an efficient buffer in helping E. coli survive in soil environments and in waterways (13). Thus, perhaps, it is not surprising that E. coli was isolated from farm lands where cattle had not grazed for three years. Fecal E. coli have also been isolated in large numbers from waterways off the shores of Great Britain, and from streams taking effluents from sewage in Chile and the United States. Recent studies from our laboratory have shown a high frequency of plasmid-bearing antibiotic resistant non-coliform and coliform Gram-negative organisms associated with common fruits and vegetables (in preparation). These findings raise concern about factors and conditions which select and enhance plasmid survival in the environment.

The majority of E. coli of fecal origin is disposed of, at least in the developed world, in a sewage system consisting of several processes. In the United States, most cities use a primary settling system with lagooning or secondary biological treatment before disinfection and discharge into surface waters or area ponds. The primary and secondary sludges may be dewatered, buried or incinerated. Occasionally, they are used as land-fill with or without further treatment. Alternatively, this sludge may be digested anaerobically and then dewatered before disposal. In order to study plasmid survival of E. coli in these systems, individuals have set up bench-scale model treatment plants, and introduced appropriately-reduced amounts of sewage. In one such model treatment plant, Sagik et al compared survival of a genetically-marked isolated sewage E. coli, GF215 with E. coli K12 (17). The raw waste-water was seeded with approximately 5×10^7 viable organisms/100ml, and sewage was sampled over 120 hours. Survival of the organisms was tested first in the raw waste water reservoir and then in the primary and secondary lagoons. This group found no difference in the decay constant for the wild-type isolate and two prototypes of parental E. coli K12. Decay under anaerobic digestion conditions at 37°, however, was considerably more rapid for the E. coli K12 organisms than for the wild-type. When they tested two different debilitated E. coli hosts, both classified as EK2 host vectors, they found a curve similar to normal E. coli K12 for Dp50supF, but more rapid disappearance of χ1776 which reached undetectable levels within 20 hours. Tests of these same strains bearing plasmids did not show any dramatic increase in survival over the non-plasmid containing strains. In general, looking at wild-type E. coli and E. coli K12, reductions of 20-30% as a result of primary treatment, and 95-99% removal as a result of secondary treatment, were found. Thus, it can be concluded that the removal of EK1 and EK2 hosts during the process of conventional sewage treatment parallels that of the waste-water bacteria. The total survival of E. coli may be to the extent of about 10^{-3}-10^{-4}. These investigators could demonstrate no transfer of pBR322 to indigenous sewage flora after 25 hours of stationary cultures at 37°.

In more recent studies, Watkins et al (personal communication) constructed a multiply-resistant poorly-mobilizable plasmid, pES019, from pBR328, put it into both wild-type and laboratory strains of E. coli K12, and introduced this into a model treatment plant (Figure 3). The results of these studies showed that E. coli K12 and wild-type E. coli host strains survived sewage treatment at nearly the same levels. No evidence for mobilization of the plasmid to bacterial strains in sewage was noted. In a separate group of experiments, these investigators seeded the primary sewage tank with 100 µg/liter of naked plasmid pES019 DNA. They found no evidence for uptake and transformation of indigenous bacteria by the DNA.

On the other hand, studies by Sagik et al, by Watkins et al and by
others (18) have shown a seeming enrichment for antibiotic resistant
strains in sewage during the treatment process. The antibiotic
resistant strains appear to survive better as a group than do
sensitive strains. Since this was not noted with the cloned small
resistance plasmids, one can surmise that survival may have to do
with other genes on the naturally-occurring resistance plasmids.

Of particular note in all these studies was the large number of
viable bacteria left in the sludge from the primary and secondary
treatment plant. Thus, the handling of this sludge, as well as the
liquid effluent should be critically evaluated before it is
dispersed into the environment.

We should be more aware of how we dispose of the organisms that are
grown. The data show that .01% of incoming organisms survive sewage
treatment unless disinfectants or dewatering are used. I make this
point because I am not as worried about environmental introduction
of recombinant molecules, (these are generally carefully handled,
debilitated, and inactivated before being put into the sewage
system) as I am about non-recombinant resistance plasmids. Their
host bacteria can be dumped into the sewage system in large
quantities with no restrictions. It is important that they also be
carefully handled and inactivated before being discarded.

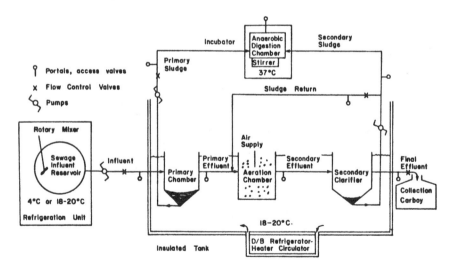

Figure 3.   Schematic of wastewater treatment model (Watkins et al).

ACKNOWLEDGEMENTS

Most of the studies quoted from my laboratory were supported by NIAID Contract NIA 72529. Use of the Clinical Study Unit of Tufts-New England Medical Center was made possible through grant 5M01RR00054-18.

REFERENCES

1. Freter, R. (1978). Possible effects of foreign DNA on pathogenic potential and intestinal proliferation of Escherichia coli. J. Inf. Dis. 127, 624-628.

2. Curtiss III, R., Inoue, M., Pereira, D., Hsu, J.C., Alexander, L. and Rock, L. (1977). Construction and use of safer bacterial host strains for recombinant DNA research. in Miami Winter Symposia, Vol. 13, Molecular Cloning of Recombinant DNA. (Scott, W.A. and Werner, R. eds) pp. 99-114, Academic Press, New York.

3. Levy, S.B., Marshall, B., Onderdonk, A. and Rowse-Eagle, D. (1980). Survival of Escherichia coli host-vector systems in the mammalian intestine. Science 209, 391-394.

4. Smith, H.W. (1975). Survival of orally administered E. coli K12 in alimentary tract of man. Nature 255, 500-502.

5. Anderson, E.S. (1975). Viability of, and transfer of a plasmid from, E. coli K12 in the human intestine. Nature 255, 502-504.

6. Marshall, B., Schluederberg, S., Tachibana, C. and Levy, S.B. (1981). Survival and transfer in the human gut of poorly mobilizable (pBR322) and of transferable plasmids from the same carrier E. coli. Gene 14, 145-154.

7. Levine, M.M., Kaper, J.B., Lockman, H., Black, R.E., Clements, M.S. and Falkow, S. Recombinant DNA risk assessment studies in man: efficacy of poorly-mobilizable plasmids in biologic containment. (manuscript submitted).

8. Levy, S.B., Fitzgerald, G.B. and Macone, A.B. (1976). Spread of antibiotic-resistance plasmids from chicken to chicken and chicken to man. Nature 260, 40-42.

9. Anderson, J.D., Gillespie, W.A. and Richmond, M.H. (1973). Chemotherapy and antibiotic resistance transfer between enterobacteria in the human gastrointestinal tract. J. Med. Microbiol. 6, 461-473.

10. Sears, H.J., Brownlee, I. and Vchiyama, J.K. (1950). Persistence of individual strains of Escherichia coli in the intestinal tract of man. J. Bacteriol. 59, 293-301.

11. Onderdonk, A., Marshall, B., Cisneros, R. and Levy, S.B. (1981). Competition between congenic Escherichia coli K12 strains in vivo. Inf. Imm. 32, 74-79.

12. Myhal, M.L., Laux, D.C. and Cohen, P.A. (1982). Relative colonizing activities of human fecal and K12 strains of Escherichia coli in the large intestines of streptomycin-treated mice. Eur. J. Clin. Microbiol. 1, 186-192.

13. Stotsky, G. and Krasowsky, V.N. (1981). Ecological factors that affect the survival, establishment, growth and genetic recombination of microbes in natural habitats. In Molecular Biology, Pathogenicity and Ecology of Bacterial Plasmids, (Levy S.B., Clowes, R.C. and Koenig, E.L. eds) pp. 31-42, Plenum Press, New York.

14. Smith, H.W. (1969). Transfer of antibiotic resistance from animal and human strains of Escherichia coli to resident E. coli in the alimentary tract of man. Lancet i, 1174-1176.

15. Petrocheilou, V., Grinsted, J. and Richmond, M.H. (1976). R plasmid transfer in vivo in the absence of antibiotic selection pressure. Antimicrob. Agents Chem. 10, 753-761.

16. Williams, P.H. (1977). Plasmid transfer in the human alimentary tract. FEMS Microb. Lett. 2, 91-95.

17. Sagik, B.P., Sorber, C.A. and Moore, B.E. (1981). The survival of EK1 and EK2 systems in sewage treatment plant models. In Molecular Biology, Pathogenicity and Ecology of Bacterial Plasmids, (Levy, S.B., Clowes, R.C. and Koenig, E.L. eds) pp. 31-42, Plenum Press, New York.

18. Grabow, W.O.K., van Zyk, M. and Prozesky, O.W. (1976). Behavior in conventional sewage purification processes of coliform bacteria with transferable or non-transferable drug-resistance. Water Res. 10, 717-723.

GENE CLONING AND BACTERIAL PATHOGENICITY

W. Goebel[1], J. Hacker[1], C. Hughes[1], S. Knapp[1], H. Hof[2],
D. Müller[1], A. Juarez[1] and J. Kreft[1]

[1] Institut für Genetik und Mikrobiologie, Universität
Würzburg, Würzburg, W. Germany

[2] Institut für Hygiene und Mikrobiologie, Universität
Würzburg, Würzburg, W. Germany

ABSTRACT

The genetic cloning technique offers the opportunity to isolate single
genetic determinants which may be involved in the often complex
picture of pathogenicity observed among bacterial species. The contribution of the isolated virulence determinant can be measured in a
suitable in vivo model system and attempts can be made to develop a
vaccine based on one or more characterized virulence properties. Two
such properties, synthesis of hemolysin and mannose-resistant
hemagglutination, which we examined by gene cloning techniques, were
identified as virulence determinants in E. coli causing extraintestinal
infections and the isolation of the genetic determinants enabled us to
unravel in part the complex biochemical processes behind these
properties. Furthermore the need for using different host/vector
systems when studying virulence properties of Gram-positive pathogenic
bacteria, is pointed out.

INTRODUCTION

Pathogenicity is the relatively rare ability of microorganisms (virus,
bacteria, fungi, protozoa) to cause disease. I shall discuss here
only pathogenicity of bacteria. The vast majority of the bacteria are
harmless or even beneficial to their human, animal or plant hosts but
this small group of pathogenic bacteria can cause a variety of
diseases ranging from relatively harmless disturbances of the infected
eucaryotic organism to severe illness and even death. Improved
hygiene and nutrition and the application of antibacterial drugs have
drastically reduced the outbreaks of devastating bacterial epidemics
in man, at least in industrialized countries, but none has been as
yet eliminated and other bacterial infections, debilitating and
economically damaging, have replaced them in importance. Since many
bacterial species of which pathogenic forms are known are often
associated as normal commensal microorganisms with man, animals or
plants, the question arises: What distinguishes "pathogenic"
bacteria from their normal "harmless" relatives?, a question which
has been asked ever since bacteria were recognized as causative agents
of infectious disease. To understand bacterial pathogenicity we should
perhaps first consider the various stages often seen during

bacterial infections or man.

1) Interaction of the bacteria with the skin of a mucuous membrane and colonization of the bacteria at these primary sites of infection.

2) Penetration through the skin or the mucuous membrane and multiplication in the host tissue. The interaction of the infecting bacteria with the tissue may be either specific or unspecific.

3) Occasionally bacteria may even invade and infect other parts of the host.

4) Interference of the penetrating or invading bacteria with the nonspecific and the immunospecific defense system of the host.

5) Damage to the host.

To successfully carry out these stages of infection, bacteria must possess specific characters, e.g. adhesion ability, toxicity and resistance to complement, macrophages and other factors of the host defense system. Such characters are determined by a range of bacterial surface antigens and extracellular products, termed pathogenicity or virulence factors, and these may be encoded by chromosomal and/or extrachromosomal genes. With the exception of a few highly toxic species, the pathogenicity of which is due almost exclusively to toxin production (Clostridium tetani, Clostridium botulinum, Cornyebacterium diphtheriae), pathogenic bacteria possess a complex array of putative pathogenicity factors which makes difficult the identification and characterization of individual components. Most earlier studies of suspected pathogenicity (virulence) factors comprised comparisons of wild-type virulent strains with avirulent isolates (or more recently avirulent variants obtained by mutagenesis) in suitable animal models. The nature of the events (natural selection, mutation) leading to the avirulent state of bacteria was in general unknown as was the genetic complexity of the virulence property under study.

The modern genetic techniques, especially DNA cloning, offers now the possibility to isolate the DNA segment which determines a putative virulence property. Subsequently, the genes required for this property and also their products can be characterized and the biochemical reactions leading to the (often complex) virulence phenotype can be studied. The purified virulence determinant can then be transferred into a genetically well-characterized avirulent bacterial host and its contribution to virulence can be measured provided suitable cell culture or animal model systems are available. The ultimate practical application of such studies can be the development of a vaccine against the pathogen with the help of the purified virulence factor.

Critical questions have been raised whether this approach may generate new forms of pathogenic microorganisms since often an apathogenic bacterial species (in many of the experiments already performed (Table 1) it was Escherichia coli) is used as host for the cloned virulence determinant. In my opinion this risk is low since - as I pointed out before - virulence is generally a multifactorial phenomenon and a single cloned virulence factor does not provide full virulence; the bacterial host harbouring a cloned virulence determinant will thus be considerably less pathogenic than the original pathogenic donor. Other problems like reduced expression of the determinant in a distantly related new host or lack of optimal physiological conditions required for the precise expression of the virulence phenotype, will further weaken virulence of the host transformed with a recombinant DNA carrying this determinant.

In the past five years such molecular analysis of pathogenicity has been successfully applied to a number of bacteria causing disease in man, animals and plants. In Table 1 some of these bacteria and the virulence determinant studied are summarized. Many more are presently under investigation in several laboratories around the world. I would like to discuss mainly our own work on hemolysin formation in E. coli and Bacillus cereus and also the hemagglutination properties of E. coli. These investigations will show in some more detail the potential of the gene cloning technique in the analysis of bacterial pathogenicity.

RESULTS AND DISCUSSION

Hemolysins are extracellular proteins synthesized by a large group of bacteria. They cause lysis of erythrocytes (hemolysis) and often many other eucaryotic cells (cytolysis). Since hemolysin production is often associated with pathogenic bacteria this phenotype has long been viewed as a virulence factor. Among the Gram-negative bacteria, hemolysin formation is frequently encountered in Escherichia coli strains which cause extraintestinal infections such as those of the urinary tract. I should point out here that Escherichia coli is one of the most frequent bacterial infections in man. It is not only the major cause if diarrhoeal diseases (enterotoxigenic and enteropathogenic E. coli) but also of many extraintestinal infections (e.g. infection of the urinary tract is caused in about 70 % of all cases by E. coli, Hughes et al., 1983). These UTI E. coli strains generally exhibit several other properties assumed to be associated with virulence and these are summarized in Table 2. Virulence factors of UTI E. coli strains are determined by chromosomal genes in contrast to the virulence properties of enterotoxigenic E. coli strains (enterotoxins, adhesins) which are in general determined by plasmids. Plasmid-encoded synthesis of hemolysin has also been demonstrated in E. coli strains isolated from animal faeces but in general not in UTI strains. (De la Cruz et al., 1980, Müller et al., 1980 . See Bull. of WHO 58: 23-36, scientifc working group reports 1978-1980).

TABLE 1.   Cloning of bacterial pathogenicity factors

A. Gram-negative bacteria

Bacteria	Virulence determinant	Involved in	Reference
Escherichia coli	Heat-stable enterotoxin (st)	Intestinal infections in man and animals	(So et al., 1976)
	Heat-labile enterotoxin (lt)		(So et al., 1978) (Yamamoto and Yokota, 1980)
	Adhesion fimbriae		
	a) K88, K99, CFAI		(Mooi et al., 1979) (Kehoe et al., 1981) (Stripley et al., 1981) (Van Emden et al., 1980) (Maas et al., un-published)
	b) Msh-pili (Common, type 1)	Extra-intestinal infections	(Hull et al., 1981) (Hacker et al., unpublished)
	Mrh-pili (HA-type V and VI)	UTI Bacteremia	(Hull et al., 1981) (Clegg, 1982) (Berger et al., 1982)
	Iron transport systems (Aerobactin, ColV)	Meningitis of Neonates etc.	(Bindereif and Neilands, 1983) (Krone et al., 1983)
	Capsules (K1-Antigen)		(Silver et al., 1981)
	Hemolysins		(Noegel et al., 1981) (Welch et al., 1981) (Goebel and Hedgpeth, 1982) (Berger et al., 1982)
	Serum resistance (traT, iss)		(Moll et al., 1980) (Binns et al., 1979)
Neisseria gonorrhoeae	Pilus IgA protease	Gonorrhoea	(Meyer et al., 1982) (Kommey et al., 1982)
Pseudomonas aeruginosa	Hemolysin (phospholipase C)	Burns and wounds, lung-infection	(Vasil et al., 1982) (Coleman et al., 1982)
Agrobacterium tumefaciens	T-DNA	Tumor formation of dicotyledons	(Leemans et al., 1982) (Knauf and Nester, 1982)

B. Gram-positive bacteria

Bacteria	Virulence determinant	Involved in	Reference
Bacillus thuringiensis	Cristalline protein (Endotoxin)	Killing of insects	(Held et al., 1982)
Bacillus cereus	Hemolysin (Cereolysin)	?	(Kreft et al., 1983)

TABLE 2. Putative virulence factors of E. coli strains
causing infections of the urinary tract

Lipopolysaccharide (O)-antigens e.g. 04, 06, 018, 075.

K-antigens e.g. K5, K12, K13, K15.

Hemagglutination antigens e.g. MrhV, MrhVI

Serum resistance by outer membrane proteins e.g. iss

Hemolysin synthesis

Original analysis of the genetics of hemolysin production was performed on one such plasmid, pHly152 (Noegel et al., 1979; Noegel et
al., 1981). By mutagenesis with transposon 3 and 5 (Tn3, Tn5) we first
identified the DNA segment on pHly152 required for the hemolytic
phenotype and showed that it consists of four genes, which are
arranged and expressed as shown in Fig. 1. The two genes hlyA and hlyB
are needed for the synthesis of active hemolysin and its subsequent
transport across the cytoplasmic membrane. This involves the processing
of the primary gene product of hlyA into several fragments by a
presumably autoproteolytic mechanism and the modification of these non-
hemolytic proteolytic fragments by the gene product of hlyC. The
nature of this modification is as yet unknown but it is a prerequiste
for obtaining hemolytic activity. While the primary gene product of
hlyA can not be transported across the cytoplasmic membrane, its
major proteolytic fragment, a 60.000 dalton protein, and presumably
several other fragments of the hlyA gene product are readily transported. For the proteolytic cleavage and the transport of the
proteolytic fragments of hlyA gene product the hlyC product does not
seem to be required. Hemolysin activity is found exclusively in the
periplasmic space when only hlyC and hlyA are present; only the
addition of the genes hlyB$_a$ and hlyB$_b$ leads to the transport of
hemolysin across the outer membrane. Thus the gene products of hlyB$_a$
and hlyB$_b$ which have been identified as outer membrane proteins
(Noegel et al., 1979; Wagner et al., 1983; Schießl, S., Härtlein, M.
and Goebel, W., unpublished results) seem to form a hemolysin-specific
transport system for the outer membrane.

Hybridization experiments were performed between the cloned genes of
this hly-determinant of plasmid pHly152 and many other hemolytic
E. coli strains which carry the hly determinants on plasmids or the
chromosome. They all showed extensive hybridization, indicating that
all chromosomal and extrachromosomal hly genes are closely related
(De la Cruz et al., 1980; Müller et al., 1983).

This has been further substantiated by the functional exchange of
corresponding chromosomal and extrachromosomal hly genes. Nevertheless when the cloned chromosomal hly determinants from hemolytic
E. coli strains belonging to the four O-serogroups most frequently
found in urinary tract infections, i.e. 04, 06, 018 and 075 were
analysed with restriction enzymes, we observed specific differences.

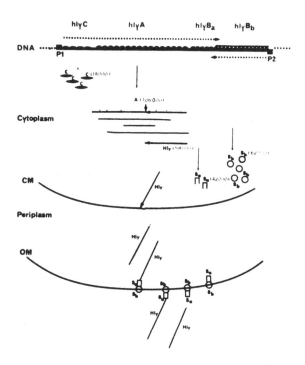

Figure 1. The arrangement and expression of the four hly
genes and a model for the interaction of the four gene products
(C, A, B$_a$ and B$_b$) in the formation of active hemolysin and its transport acróss the cytoplasmic (CM) and the outer membrane (OM). P1 and
P2 indicate the two promoters from where the four hly genes are
transcribed. Broken lines indicate the transcripts. For further
details see text.

These variations seem to be preferentially located in hlyA, the gene
which has been shown to determine the hemolytically active product(s),
indicating that the hemolysins produced by these strains may not be
completely identical (Berger et al., 1982).

Hemolysin production in E. coli is not a completely stable phenotype. This is readily explicable for the extrachromosomal type, but even hemolytic E. coli strains which carry the hly determinant on the chromosome yield non-hemolytic mutants at an unexpectedly high rate (10^{-3} to 10^{-4}). In one 06 strain most of these Hly⁻ mutants have lost in addition to the hemolytic phenotpye the mannose-resistant hemagglutination (Mrh) character (Hacker et al., 1983). This property is connected with the formation of specific protein pili which are present on most E. coli strains isolated from urinary tract infections and which are assumed to cause specific adhesion of the bacteria to epithelial cells (Svanborg-Eden et al., 1977) of the urinary tract. In addition to these Mrh-pili, most E. coli strains form the "common " (type I) pili which cause agglutination of guinea pig erythrocytes sensitive to mannose (Klemm et al., 1982). The mrh determinant (mrhVb) of this E. coli 06 strain was first isolated on a recombinant cosmid which was identified in a gene bank obtained from the chromosomal DNA of this strain. Subcloning of the insert of this cosmid clone yielded two recombinant plasmids. One of them caused mannose-resistant hemagglutination of bovine erythro-cytes, as expected for mrhVb, and the other caused mannose-sensitive hemagglutination of guinea pig erythrocytes characterisitic for "common" pili (type I). Restriction enzmye analyses and hybridization of these recombinant DNAs demonstrated extensive similarities between both recombinant plasmids. With the cloned hly- and mrh- determinants we have analysed the previously described Hly⁻, Mrh⁻ mutants by hybridization of their chromosomal DNA with the corresponding radioactively labeled recombinant plasmids and have found that all Hly⁻, Mrh⁻ mutants have suffered similar extended deletions which seem to be triggered by IS-like elements flanking both determinants (Fig. 2). These mutants thus represent genetically well-characterized hosts for the study of the virulence properties of the isolated hly and mrh determinants. The four hly determinants studied in the following experiments were cloned in such a way that the genetic environment excludes the possibility that functional differences may be caused by unequal expression of the hemolysin genes due to differing transcription signals.

The cloned putative virulence determinants under study were trans-formed into the same Hly⁻, Mrh⁻ mutant and transformed strains were applied at given doses intraperitoneally to inbread mice (Hacker, J., C. Hughes, H. Hof and W. Goebel, submitted for publication). The lethality of the mutant E. coli strains (Hly⁻, Mrh⁻) compared to the wild-type isolate (Hly⁺, Mrh⁺) is clearly reduced (data not shown). Reintroduction of the three cloned chromosomal hly determinants, but not of the cloned extrachromosomal hly determinant, leads to a significant increase in the toxicity of the bacteria (see Fig. 3). Similar results have been also obtained by Welch et al. using a different animal model. Our results show in addition that the chromosomal hly determinants exhibit remarkable differences in their toxic effects, i.e. the hly determinant from the 06 strain deter-mines significantly less toxicity than those from the 018 and 075 strains (Fig. 3).

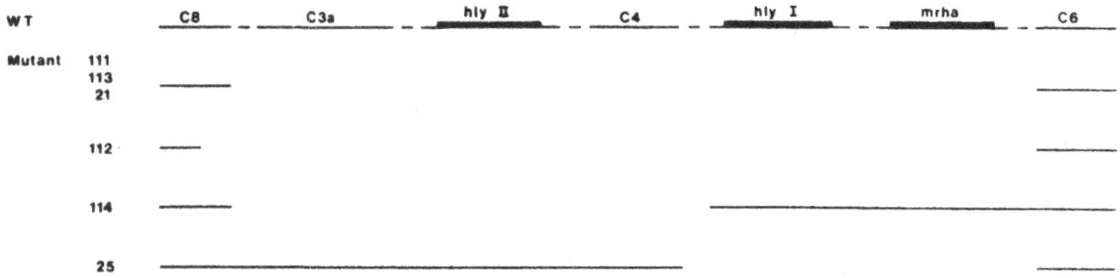

W T		C8	C3a	hly II	C4	hly I	mrha	C6
Mutant	111 113 21							
	112							
	114							
	25							

Figure 2. Arrangement of the two hly determinants (hlyI and hlyII) and the determinant for mannose-resistant hemagglutination (mrh) on the chromosome of the O6 E. coli strain 536 and spontaneous deletion mutants arising from this strain. C8, C3a, C4 and C6 indicate restriction fragments on which IS-like elements have been identified (Hacker et al., 1983) and which seem to act as start points for the deletions.

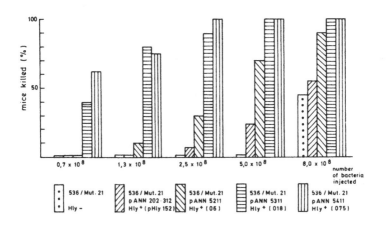

Figure 3. Lethality of E. coli strains transformed with various chromosomal and plasmid hly determinants.

No toxic effect is exerted by the mutant (Hly⁻, Mrh⁻) when a recombinant cosmid (or the subcloned plasmid) carrying the mrh
determinant was introduced. A different effect was observed when
these E. coli strains were injected intravenously and the subsequent
colonization of the bacteria in the mouse kidneys were measured
after 1 to 8 hrs post infection. As shown in Fig. 4 no significant
colonization of the bacteria is observed when the mutant strain
harbouring the various hly determinants was used. In contrast, transformation of the mrh determinant into the mutant strain results in
colonization of the bacteria in the kidney to an extent comparable
to that of the wild-type strain.

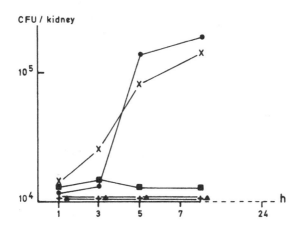

Figure 4. Colonization of E. coli strains in the kidneys of
mice carrying cloned hly or mrh determinants. CFU = colony forming
units (viable cells)

E. coli 536	●——●
536/Mut21	■——■
536/Mut21 pANN5211 (Hly⁺06)	+——+
536/Mut21 pANN5311 (Hly⁺018)	▲——▲
536/Mut21 pAN801 (Mrh⁺Vb)	✕——✕

These results confirm experimentally that both the hly and the mrh
determinants code for properties which are directly involved in pathogenicity but in different steps of the infection. The data further
indicate that a given phenotype, like hemolysis, does not automatically
predict the level of toxicity of virulence. Small variations in the

genotype may be decisive.

The experiments described up to now were performed in the same
bacterial species from which the putative virulence factors originate.
Possible biohazards arising from the cloning of virulence properties
into E. coli are therefore insignificant. In the following I would
like to address this problem by describing the cloning of a toxin
gene in a heterologous host, i.e. the cloning of the cereolysin gene
into E. coli and Bacillus subtilis.

Cereolysin is produced by Bacillus cereus strains. It is a cytotoxin
which acts on all animal cells and will kill mice when injected at a
dose of < 100 ug/mouse (Gill, 1982). Several other cytolytic toxins
produced by Bacillus and Clostridium species, by pyogenic Streptococci
(streptolysin O) and by Listeria monocytogenes (listeriolysin) have
similar biochemical and physical properties to cereolysin. Since the
gene(s) for cereolysin synthesis are most likely chromosomally inherited the chromosomal DNA of Bacillus cereus was cloned in E. coli
K-12 by a shotgun experiment using again the cosmid cloning technique.
Among several thousand colonies tested five were found which formed
a small zone of hemolysis. The insert of the isolated recombinant
cosmid was further subloned on a shuttle vector which replicates in
E. coli and B. subtilis.

It was found that transformed E. coli again formed small hemolysis
zones on erythrocyte-agar plates whereas B. subtilis transformed
with the same recombinant DNA produced large amounts of a hemolysin
which by several biochemical and immunological tests proved to be
cereolysin (Kreft et al., 1983).

E. coli K-12 harboring the recombinant DNA with the cereolysin gene
was injected intraperitoneally into mice. As indicated in Table 3

TABLE 3.   Toxicity for mice of E. coli K-12 strains which
contain the cloned cereolysin determinant

Strain	Cereolysin-production	Bacteria injected			
		2×10^8	4×10^8	8×10^8	2×10^9
E. coli K-12	-	$10/0$[1]	$10/0$	$10/3$	$10/9$
F. coli K-12 pJKK1-hlyl	+	$10/0$	$10/0$	$10/4$	$10/9$

[1] mice injected/mice killed

there was no increase in toxicity of the E. coli transformants
compared to the E. coli K-12 host even at high titers of injected
bacteria. This indicates that the cereolysin gene, despite being
expressed in E. coli does not significantly increase "virulence"
of the K-12 strains although isolated cereolysin exhibits
toxicity in mice even at a rather low dosage.

ACKNOWLEDGEMENTS

This work was supported by the Deutsche Forschungsgemeinschaft
(SFB 105, A-12 and Go 168/11-2). We thank M. Vogel and E. Appel
for help in the preparation of the manuscript.

REFERENCES

Berger, H., Hacker, J., Juarez, A., Hughes, C. and Goebel, W. (1982).
    Cloning of the chromosomal determinants encoding hemolysin
    production and mannose-resistant hemagglutination in
    Escherichia. coli. J. Bacteriol. 152, 1241-1247

Bindereif, A. and Neilands, J.B. (1983). Cloning of the aerobactin-
    mediated iron assimilation system of plasmid ColV.
    J. Bacteriol. 153, 1111-1113

Binns, M.M., D.L. Davis and Hardy, K.G. (1979). Cloned fragments of
    the plasmid ColV, I-K94 specifying virulence and serum
    resistance. Nature 279, 778-781

Clegg, S. (1982). Cloning of genes determining the production of
    mannose-resistant fimbriae in a uropathogenic strain of
    Escherichia coli belonging to serogroup O6. Infect. Immun.
    38, 739-744

Coleman, K., Dougan, G. and Arbuthnott, J.P. (1983). Cloning, and
    expression in Escherichia coli K-12, of the chromosomal
    hemolysin (phospholipase C) determinant from Pseudomonas
    aeruginosa. J. Bacteriol. 153, 909-915

De la Cruz, F., Müller, D., Ortiz, J.M. and Goebel, W. (1980). A
    hemolysis determinant common to Escherichia coli Hly-
    plasmids of different incompatibility groups. J. Bacteriol.
    143, 825-833

Gill, D.M. (1982). Bacterial toxins: a table of lethal amounts.
    Microbiol. Rev. 46, 86-94

Goebel, W. and Hedgpeth, J. (1982). Cloning and functional characte-
    rization of the plasmid-encoded hemolysin determinant of
    Escherichia coli. J. Bacteriol. 151, 1290-1298

Hacker, J., Knapp, S. and Goebel, W. (1983). Spontaneous deletions
    and flanking regions of the chromosomally inherited hemo-
    lysin determinant of an Escherichia coli O6 strain.
    J. Bacteriol. 154, in press

Held, G.A., Bulla, L.A., Ferrari, E., Hoch, J. and Aronson, A.J. (1982).
    Cloning and localization of the lepidopteran protein gene of
    Bacillus thuringiensis subsp. Kurstaki. Proc. Natl. Acad.
    Sci. U.S.A. 79, 6065-6073

Hughes, C., Hacker, J., Roberts, A. and Goebel, W. (1983). Hemolysin
    production as a virulence marker in symptomatic and
    asymptomatic urinary tract infections caused by Escherichia
    coli. Infect. Immun. 39, 546-551

Hull, R.A., Gill, R.E., Hsu, P., Minshew, B.H. and Falkow, S. (1981).
    Construction and expression of recombinant plasmids encoding
    type 1 or D-mannose-resistant pili from urinary tract in-
    fection Escherichia coli isolate. Infect. Immun. 33, 933-938

Kehoe, M., Sellwood, R., Shipley, P. and Dougan, G. (1981). Genetic
    analysis of K88-mediated adhesion of enterotoxigenic
    Escherichia coli. Nature, 291, 122-126

Klemm, P., Ørskov, J. and Ørskov, F. (1982). F7 and type 1 like
    fimbriae from three Escherichia coli strains isolated from
    urinary tract infections: protein chemical and immunological
    aspects. Infect. Immun. 36, 462-468

Knauf, V.C. and Nester, E.W. (1982). Wide host range cloning vectors:
    a cosmid clone bank of an Agrobacterium Ti plasmid. Plasmid
    8, 45-54

Koomey, J.M., Gill, R.E. and Falkow, S. (1982). Genetic and biochemi-
    cal analysis of gonococcal IgA1 protease: Cloning in
    Escherichia coli and construction of mutants of gonococci
    that fail to produce the activity. Proc. Natl. Acad. Sci.
    U.S.A. 79, 7881-7883

Kreft, J., Berger, H., Härtlein, M., Müller, B., Weidinger, G., and
    Goebel, W. (1983). Cloning and expression of the hemolysin
    (cereolysin) determinant from Bacillus cereus in
    Escherichia coli and Bacillus subtilis. J. Bacteriol., 154,
    in press

Krone, W.J.A., Oudega, B., Stegehuis, F. and De Graaf, F.K. (1983).
    Cloning and expression of the cloacin DF 13/aerobactin
    receptor of Escherichia coli (ColV-K30). J. Bacteriol. 153,
    716-721

Leemans, K., Deblaere, R., Willnitzer, L., De Greve, H., Hernolsteens, J., Van Montagu, M. and Schell, J. (1982). Genetic identification of functions of Ti-DNA transcripts in octopine crown galls. EMBO J., 1, 147-152

Meyer, T.F., Mlaver, N. and So. M. (1982). Pilus expression in Neisseria gonorrhoeae involves chromosomal rearrangements. Cell 30, 45-54

Moll, A., Manning, P.A. and Timmis, K.N. (1980). Plasmid-determined resistance to serum bactericidal activity: a major outer membrane protein the traT gene product, is responsible for plasmid-specified serum resistance in Escherichia coli. Infect. Immun. 28, 359-367

Mooi, F.R., De Graaf, F.K., and Van Embden, J.D.A. (1979). Cloning, mapping and expression of the genetic determinant that encodes for the K88 ab antigen. Acids Res. 6, 849-865

Müller, D., Hughes, C. and Goebel, W. (1983). Relationship between plasmid and chromosomal hemolysin determinants of Escherichia coli. J. Bacteriol. 153, 846-851

Noegel, A., Rdest, U. and Goebel, W. (1981). Determination of the functions of hemolytic plasmid pHly152 of Escherichia coli. J. Bacteriol. 145, 233-247

Noegel, A., Rdest, U., Springer, W. and Goebel, W. (1979). Plasmid cistrons controlling synthesis and excretion of the exotoxin α-hemolysin of Escherichia coli. Molec. Gen. Genet. 175, 343-350

Shipley, P.L., Dougan, G. and Falkow, S. (1981). Identification and cloning of the genetic determinant that encodes for the K88 ac adherence antigen. J. Bacteriol. 145, 920-925

Silver, R.P., Finn, C.W., Vann, W.F., Aaronson, W., Schneerson, R., Kretschmer, P.J. and Garon, C.F. (1981). Molecular cloning of the K1 psular polysaccharide gene of E. coli. Nature 289, 696-698

So, M., Boyer, H.W., Betlach, M. and Falkow, W. (1976). Molecular cloning of an Escherichia coli plasmid determinant that encodes for the production of heat-stable enterotoxin. J. Bacteriol. 128, 463-472

So, M., Dallas, W.S. and Falkow, S. (1978). Characterization of an Escherichia coli plasmid encoding for synthesis of heat-labile toxin: molecular cloning of the toxin determinant. Infect. Immun. 21, 405-411

Svanborg, Edén, C., Erikson, B. and Hanson, A.A. (1977). Adhesion
    of Escherichia coli to human uroepithelial cells in vitro.
    Infect. Immun. 18, 767-774

Van Embden, J.D.A., De Graaf, F.K., Schouls, L.M. and Teppema, J.S.
    (1980). Cloning and expression of a deoxyribonucleic
    acid fragment that encodes for the adhesive antigen K99.
    Infect. Immun. 29, 1125-1133

Vasil, M.L., Berka, R.M., Gray, G.L. and Nakai, H. (1982). Cloning
    of a phosphate-regulated hemolysin gene (phospholipase C)
    from Pseudomonas aeruginosa. J. Bacteriol. 152, 431-440

Wagner, W., Vogel, M. and Goebel, W. (1983). Transport of hemolysin
    across the outer membrane of Escherichia coli requires two
    functions. J. Bacteriol. 154, in press

Welch, R.A., Dellinger, E.P., Minshew, B. and Falkow, S. (1981).
    Hemolysin contributes to virulence of extraintestinal
    Escherichia coli infections. Nature 294, 665-667

Yamamoto, T. and Yokota, T. (1980). Cloning and deoxyribonucleic acid
    regions encoding a heat-labile and heat-stabile enterotoxin
    originating from an enterotoxigenic Escherichia coli strain
    of human origin. J. Bacteriol. 143, 652-660

DISCUSSION

R. ROHL: What is the selective advantage for E. coli to develop the
hemolysin system, or in general, what is the selective advantage for a
microorganism to develop pathogenic agents?

W. GOEBEL: This is a difficult question to answer and I can offer you
only speculations. In general one can probably say that pathogenic
microorganisms can colonize ecological niches which are not easily
accessible to their "apathogenic" relatives. The pathogenicity
(virulence) factors will help the microorganism to accomplish this.
With regard to hemolysin, one could think that this agent, due to its
cytolytic activity; may lyse target cells and provide the
microorganism with nutrients, e.g. Fe^{2+}. Another possibility is the
inactivation of part of the host's defense system. It has been shown
in vitro that the hemolysin of E. coli can lyse leukocytes.

SELECTIVE EVOLUTION OF GENES FOR ENHANCED DEGRADATION OF
PERSISTENT, TOXIC CHEMICALS

A. M. Chakrabarty, J. S. Karns, J. J. Kilbane and D. K.
Chatterjee

Department of Microbiology, University of Illinois Medical
Center, Chicago, Illinois 60612, U.S.A.

ABSTRACT

The role of degradative plasmids in the evolution of new catabolic
capabilities has been discussed with relation to the biodegradation
of chlorinated benzoic acids. Evolution of a set of genes encoding
complete degradation of a normally persistent compound such as
2,4,5-trichlorophenoxyacetic acid has been achieved under strong
selective conditions in the chemostat. Studies concerning the
substrate specificity and regulation of the degradative enzymes
demonstrate the potential of such laboratory-developed strains to
degrade a variety of halogenated compounds for their effective
removal from the environment.

INTRODUCTION

In the industrialized world, chemicals play a vital role in our day
to day life and in sustaining agricultural productivity, eradicating
many forms of infectious diseases and in maintaining a sound, robust
economy. Agricultural chemicals, such as Dichlorodiphenyltrichloro-
ethane (DDT), for example, have played a major role not only in
enhanced agricultural productivity, but also in eradicating diseases
such as malaria. Highly chlorinated compounds such as DDT, however,
do not occur naturally, and are introduced into the environment in
large quantities as herbicides and pesticides. Because such
compounds are synthetic and in general are toxic to insects and
pests, they often accumulate in nature because of an extremely low
rate of biodegradation by natural microbial flora. Such persistence
has resulted in the contamination of human beings through their
entry via the food chain. This in turn has resulted in the
curtailment of production and use of many such chemicals as DDT,
polychlorinated biphenyls (PCBs), 2,4,5-trichlorophenoxyacetic acid
(2,4,5-T), etc. Since the chemical industry has continued, and is
likely to continue, manufacturing highly halogenated compounds for
use as agricultural chemicals, and since natural microorganisms do
not appear to cope with the multitude of such chemicals released
into the environment in massive amounts, some mechanism must be
found that would allow enhanced biodegradation of highly halogenated
compounds by natural microorganisms. In this article, we discuss
some of the genetic mechanisms that allow a 3-chlorobenzoate (3Cba)

degrading strain of Pseudomonas putida to utilize different and
higher chlorinated forms such as 4-chlorobenzoate (4Cba) or
3,5-dichlorobenzoate (3,5-Dcb). An understanding of the genetic
mode of biodegradation of such compounds has enabled us to develop a
highly selective method that has allowed the selective evolution of
the degradative capability against a known persistent compound such
as 2,4,5-T. The selective methods and some of the properties of the
enzymes involved in the dehalogenation of halobenzoates, 2,4,5-T or
chlorophenols are described in this article.

Genetic rearrangements and evolution of new degradative functions.
The 3-chlorobenzoate degradative genes in P. putida AC858 are borne
on a transmissible plasmid pAC25 which is about 117 kilobase pairs
(kb) in size (Chatterjee and Chakrabarty, 1983). The presence of
this plasmid, however, does not allow the host cells to utilize
4-chloro- or 3,5-dichlorobenzoate. It is possible to select for
4Cba$^+$ character, provided the $_{pAC25}+$ cells are grown in a chemostat
with cells harboring the TOL plasmid during enrichment with 4Cba as
a major source of carbon and energy (Chatterjee and Chakrabarty,
1982). It is known that the $_{3Cba}+$ cells cannot utilize 4Cba since
the plasmid-specified 3Cba oxygenase is not active with 4Cba;
however, the product of the oxygenase reaction, 4-chlorocatechol,
can be effectively oxidized by pAC25-coded enzymes. TOL encodes a
broad substrate specific benzoate oxygenase which is also active on
4Cba. The presence of the TOL plasmid during enrichment with 4Cba
therefore allows the transposable TOL* segment (Chakrabarty et al.,
1978) to be transposed onto the chromosome of the $_{4Cba}+$ cells (Fig.
1, top). There is usually a deletion in the pAC25 plasmid to
generate the shorter plasmid pAC27 (110 kb), which retains all the
3Cba degradative genes. It is possible to select, by combined
growth of $_{4Cba}+$ cells and cells harboring the TOL plasmid in a
chemostat in presence of 3,5-Dcb as a major source of carbon and
energy, variant cells capable of growing slowly with 3,5-Dcb.
Isolation of plasmid DNA from such cells demonstrates the presence
of both pAC27 and a second plasmid (pAC29) which has recruited the
replication/incompatibility genes of TOL and duplicate copies of a
portion of the pAC27 plasmid with mutational alterations and other
divergence in the chlorobenzoate degradative genes (Fig. 1, bottom).
Cells harboring both pAC27 and pAC29 plasmids are perfectly capable
of growing with 3Cba, 4Cba and slowly with 3,5-Dcb. Introduction of
the SAL plasmid, which is incompatible with TOL, leads to a loss of
the second plasmid (pAC29) with simultaneuous loss of the ability to
utilize 3,5-Dcb but not 3Cba or 4Cba. It is thus clear that the new
degradative genes allowing the cells to utilize 3,5-Dcb evolved as a
plasmid by recruitment of the TOL replication genes and a fragment
of the pAC27 plasmid with selective genetic divergence (Chatterjee
and Chakrabarty, 1982). It is interesting to note that TOL performs
two basic functions in such an evolutionary process: (i) it
supplies the broad substrate specific benzoate oxygenase gene as
part of the TOL* transposon on the chromosome that allows conversion

of 4Cba or 3,5-Dcb to the corresponding chlorocatechols and (ii) it
supplies the rep/inc genes that allow the replication/maintenance of
the newly evolved plasmid pAC29 with altered chlorobenzoate genes
specifying utilization of 3,5-Dcb (Chatterjee and Chakrabarty, 1982).

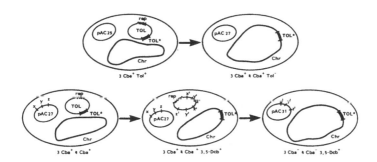

Figure 1. Diagrammatic representation of the genetic
rearrangements occurring in cells during selection with
4-chlorobenzoate (top) and 3,5-dichlorobenzoate (bottom). TOL*
represents the transposable toluate degradative gene sequence, rep
the replication gene(s), and chr the chromosome of the cell (for
details, see Chatterjee and Chakrabarty, 1982).

Selective evolution of new degradative functions. While the
evolution of pAC29 is an interesting example of the role of plasmids
in extending the degradative capability of bacteria, there are other
examples where different plasmids are involved in allowing total
degradation of other chlorinated compounds. For example, we have
demonstrated (Furukawa and Chakrabarty, 1982) that a combination of
two plasmids, pKF1 and pAC27 or pAC31 will allow total degradation
of mono- or dichlorobiphenyls. In addition, plasmids also serve as
the gene pools for the evolution of other plasmids. Thus plasmids
such as pAC25, SAL or TOL demonstrate considerable homology with
both degradative and antibiotic resistance plasmids, signifying
evolutionary relationships among one another (Bayley et al., 1979;
Kellogg et al., 1981). Based on the premise that degradative
plasmids serve as gene pools for the evolution of new degradative
capabilities in bacteria, a method has been described (Kellogg et

al., 1981) that has allowed the development initially of a mixed culture capable of utilizing as a sole source of carbon and energy a rather persistent compound such as 2,4,5-trichlorophenoxyacetic acid (2,4,5-T). Further selection with 2,4,5-T as a sole source of carbon has led to the isolation of a pure culture of P. cepacia AC1100 that can utilize not only 2,4,5-T rapidly, but can oxidize a number of chlorophenols (Kilbane et al., 1982). A schematic diagram of the selective development of this strain is given in Fig. 2.

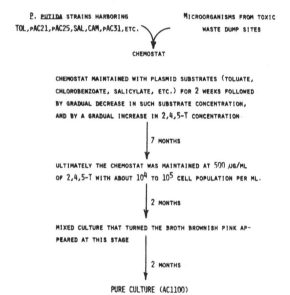

Figure 2.   Schematic representation of the method (PAMB) used for developing a 2,4,5-T degrading culture under chemostatic selective conditions (Kellogg et al., 1981; Kilbane et al., 1982).

Substrate specificity and regulation of the gene products.   Normally 2,4,5-T is oxidized slowly in nature by mixed cultures through co-oxidative metabolism (Rosenberg and Alexander, 1980). Such co-oxidation does not allow incorporation of the substrate carbons into the bacterial cell mass and therefore does not support bacterial growth (Alexander, 1981). The development of a pure culture of P. cepacia capable of utilizing 2,4,5-T as a sole carbon and energy source is therefore of considerable interest in studying the biochemistry and regulation of the degradation of this compound.

We previously demonstrated that AC1100 can not only utilize 2,4,5-T but can oxidize a number of chlorophenols (Kilbane et al., 1982). The ability of P. cepacia AC1100 to dechlorinate a variety of chlorophenols is shown in Table 1.

Table 1. Effectiveness of various chlorophenols as substrates by Pseudomonas cepacia AC1100

Substrate	Uptake (%)	Dechlorination (%)
Pentachlorophenol	100	92
2,3,4,5-Tetrachlorophenol	100	60
2,3,4,6-Tetrachlorophenol	100	80
2,3,5,6-Tetrachlorophenol	97	94
2,4,5-Trichlorophenol	100	90
2,4,6-Trichlorophenol	91	56
3,4,5-Trichlorophenol	51	27
2,3-Dichlorophenol	100	65
2,4-Dichlorophenol	100	71
2,5-Dichlorophenol	100	81
2,6-Dichlorophenol	100	36

All substrates were used at 0.1 mM. The percent uptake is represented as percent of the substrate removed from the broth in 3 hours by resting cells of AC1100. The total percent dechlorination has been measured during a 3 hour incubation with the resting cells of AC1100.

Gas chromatography—mass spectrometric studies of the intermediates isolated from the broth of AC1100 cells grown with 2,4,5-T have demonstrated a number of possible intermediates of 2,4,5-T degradation, viz. 2,4,5-trichlorophenol (2,4,5-TCP), 3,5-dichlorocatechol, mono hydroxy 2,4,5-T and mono hydroxy 4,5-dichlorocatechol (I. S. You, J. S. Karns and A. M. Chakrabarty, in preparation). That 2,4,5-TCP is a primary intermediate of 2,4,5-T metabolism has been demonstrated by following the accumulation of this compound, when resting cells of AC1100 are treated with 2,4,5-T in the presence of 50 ug/ml of chlorophenols such as 2,3,5-TCP or 2,3,4,5-tetrachlorophenol (J.S. Karns, S. Duttagupta, and A. M. Chakrabarty, in preparation). AC1100 cells also are perfectly capable of growing with 2,4,5-TCP as a sole source of carbon and energy. Studies concerning the expression of the degradative genes for 2,4,5-T and 2,4,5-TCP, when AC1100 cells

are grown with alternate sources of carbon such as succinate or
glucose, have demonstrated that the 2,4,5-T degradative enzymes are
inducible. Such enzymes are induced only when the cells are grown
with 2,4,5-T or 2,4,5-TCP. Succinate grown resting cells of AC1100,
however, can convert 2,4,5-T to 2,4,5-TCP, suggesting that the
enzyme(s) which converts 2,4,5-T to 2,4,5-TCP is constitutively
expressed (J. S. Karns and A. M. Chakrabarty, in preparation).

Since 2,4,5-TCP is an extremely toxic compound, it was of interest
to see if presence of 2,4,5-TCP would feed back inhibit the
enzyme(s) catalyzing the conversion of 2,4,5-T to 2,4,5-TCP. The
results in Fig. 3 clearly demonstrate that whereas the resting cells
of AC1100 can quickly dechlorinate 2,4,5-TCP (25 ug/ml) when it is
present as the only substrate, there is an accumulation of
increasing amounts of 2,4,5-TCP when such cells are subjected to
both 2,4,5-T (500 ug/ml) and 2,4,5-TCP (25 ug/ml). 2,4,5-T alone
was dechlorinated under these conditions without accumulation of any
2,4,5-TCP. These results clearly suggest that 2,4,5-TCP not only
fails to exert any feed back inhibition on its formation from
2,4,5-T but actually exerts toxic inhibitory effects on its
subsequent metabolism, when present at a high concentration (above
25 ug/ml). The inducibility of the 2,4,5-TCP degradative enzymes
and the constitutivity of the enzyme(s) catalyzing conversion of
2,4,5-T to 2,4,5-TCP result in growth inhibition and poisoning of
the cells, when a high concentration of 2,4,5-T (1 mg/ml) is added
to AC1100 cells growing on succinate, leading to sudden accumulation
of toxic amounts of 2,4,5-TCP.

An important aspect of the biodegradation capability of chlorobenzoate or 2,4,5-T degrading strains of Pseudomonas species is the range
of substrates that can be effectively degraded. Not only
chlorinated aromatics, but a large number of other halogenated
aromatics (polybrominated biphenyls, for example) have been released
into the environment in large amounts during the last several
decades. It was therefore of interest to us to determine the range
of halogenated compounds that can be dehalogenated by the
chlorobenzoate or 2,4,5-T and chlorophenol degrading strains. The
results in Table 2 clearly demonstrate that fluorinated and
brominated analogues are good substrates, while the iodinated
analogue of chlorobenzoic acid is a poor substrate. The iodobenzoic
acid is not only a poor inducer of the degradative enzymes for
halogenated benzoic acids but appears to inhibit induction or
dechlorination of chlorobenzoic acid. Nevertheless, it appears that
development of strains capable of utilizing highly chlorinated
compounds should in general be expected to dehalogenate the
corresponding fluoro- or bromo- analogues of these compounds.

2, 4, 5 – T (µg/ml)

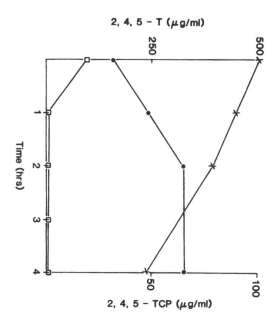

2, 4, 5 – TCP (µg/ml)

Figure 3. Production of 2,4,5-TCP from 2,4,5-T in presence of high concentrations (25 ug/ml) of 2,4,5-TCP by resting cells of P. cepacia AC1100. Such cells were incubated with 500 ug/ml of 2,4,5-T and 25 ug/ml of 2,4,5-TCP, and at various times aliquots were taken and the level of 2,4,5-TCP (●) determined by high performance liquid chromatography. 2,4,5-T (500 ug/ml) and 2,4,5-TCP (25 ug/ml) metabolism was also measured separately (X, 2,4,5-T degradation; □ , 2,4,5-TCP degradation).

Concluding remarks. In this article, we have tried to delineate the mechanisms that govern bacterial evolution of some degradative functions, particularly with relevance to that of chlorinated aromatic compounds. Plasmids appear to play a vital role in such evolutionary processes, mainly by providing the gene pools that may undergo genetic duplication and divergence to provide new enzymatic activities. Such involvement of plasmids has led to the development of a specific culture capable of utilizing a normally persistent compound such as 2,4,5-T. Such a culture is also capable of dechlorinating substantially a number of chlorophenols such as penta-, tetra-, tri- and dichlorophenols. In addition, the substrate specificities of the degradative enzymes are sufficiently broad so that analogous brominated or fluorinated compounds are also appreciably dehalogenated. This may allow the removal of such compounds from the environment.

Table 2. Dehalogenation of various substrates by P. putida
AC858 and P. cepacia AC1100

| Substrate | Bacteria | Percent Biodegradation | |
		halogen release	HPLC
3-Chlorobenzoic acid	AC858	100	99
3-Bromobenozic acid	"	100	95
3-Fluorobenzoic acid	"	100	99
3-Iodobenzoic acid	"	16	54
2,4,5-Trichlorophenoxy-acetic acid	AC1100	83	98
2,4-Dichlorophenol	"	85	92
2,4-Dibromophenol	"	60	85
Pentachlorophenol	"	94	100
Pentabromophenol	"	34	100
Pentafluorophenol	"	94	100

Extent of biodegradation has been measured both by the
release of the halogen as well as by determination of the physical
loss of the compound by high performance liquid chromatography
(HPLC).

It should be emphasized that a strain such as AC1100, which has
been developed under strong selective pressure in a chemostat with
2,4,5-T as a major carbon source, is very efficient in removing
large quantities of this compound from contaminated soil (Chatterjee
et al., 1982). The strain is known to be able to reduce 2,4,5-
T concentration from very high concentrations (10,000 to 20,000
ppm) to about 500 to 1,500 ppm in 6 weeks. The soil, once cleaned
up after AC1100 treatment, is capable of supporting the growth
of plants, which were normally inhibited by as low as 15 to 20
ppm of 2,4,5-T (Kilbane et al., 1983). Further, once the 2,4,5-
T is consumed, the strain is unable to compete effectively with
indigenous microflora and dies within a few weeks (Kilbane et al.,
1983). It is thus likely that selective evolution of degradative
genes against other normal persistent toxic chemicals may allow
ultimate removal of the compounds from the environment.

ACKNOWLEDGEMENTS

This research was supported by a grant from the National
Science Foundation (PCM81-13558), Petrogen, Inc., and by Reproductive
Hazards in the Workplace Research Grant No. 15-2 from the March
of Dimes Birth Defects Foundation. J. Karns and J.J. Kilbane acknowledge
the support of National Research Service Awards (1F 32 ES05189-
02 and 5F 32 GM08885-02 respectively).

REFERENCES

Alexander, M. (1981) Biodegradaion of chemicals of environmental
    concern. Science 211, 132-138.

Bayley, S.A., Morris, D.W. and Broda, P. (1979). The relationship of
    degradative and resistance plasmids of Pseudomonas belonging
    to the same incompatibility group. Nature 280, 338-339.

Chakrabarty, A.M., Friello, D.A. and Bopp, L.H. (1978). Transposition
    of plasmid DNA segments specifying hydrocarbon degradation
    and their expression in various microorganisms. Proc. Natl.
    Acad. Sci. U.S.A. 75, 3109-3112.

Chatterjee, D.K. and Chakrabarty, A.M. (1982). Genetic rearrangements
    in plasmids specifying total degradation of chlorinated
    benzoic acids. Mol. Gen. Genet. 188, 279-285.

Chatterjee, D.K. and Chakrabarty, A.M. (1983). Genetic homology
    between independently isolated chlorobenzoate-degradative
    plasmids. J. Bacteriol. 153, 532-534.
Furukawa, K. and Chakrabarty, D.K. (1982). Involvement of plasmids in
    total degradation of chlorinated biphenyls. Appl. Environ.
    Microbiol. 44, 619-626.

Kellogg, S.T., Chatterjee, D.K. and Chakrabarty, A.M. (1981).
    Plasmid-assisted molecular breeding: new technique for
    enhanced biodegradation of persistent toxic chemicals.
    Science 214, 1133-1135.

Kilbane, J.J., Chatterjee, D.K., Karns, J.S., Kellogg, S.T. and
    Chakrabarty, A.M. (1982). Biodegradation of 2,4,5-
    trichlorophenoxyacetic acid by a pure culture of Pseudomonas
    cepacia. Appl. Environ. Microbiol. 44, 72-78.

Kilbane, J.J., Chatterjee, D.K. and Chakrabarty, A.M. (1983).
    Detoxification of 2,4,5-T from contaminated soil by
    Pseudomonas cepacia. Appl. Environ. Microbiol. (in press).

Rosenberg, A. and Alexander, M. (1980). 2,4,5-Trichlorophenoxyacetic
    acid (2,4,5-T) decomposition in tropical soil and its co-
    metabolism by bacteria in vitro. J. Agric. Food Chem. 28,
    705-709.

DISCUSSION

S.D. EHRLICH: Why did pseudomonads develop the capacity to metabolize bizarre compounds? What is peculiar about pseudomonads? How about other bacteria?

A.M. CHAKRABARTY: I cannot answer your question as to why pseudomonads developed their extreme nutritional versatility through plasmid-mediated oxidation of exotic organic compounds, but I will try to answer the question as to how they do it. From several examples such as transfer of SAL, TOL or chlorobenzoate degradative genes to E. coli, Klebsiella, etc., we know that Pseudomonas plasmid degradative genes are poorly expressed in enteric and other bacteria, including soil bacteria. This inability to express degradative genes is believed to be due to the inability of the RNA polymerases of these bacteria to act efficiently on Pseudomonas promoter sequences. Thus, even though such plasmids are transferred to non-pseudomonal neighbors in soil, they are not properly expressed and therefore these soil microorganisms cannot degrade these compounds. But why such plasmids evolved in pseudomonads and not in other bacteria, is, however, not known.

P. DAY: You know that in agriculture, herbicide-contaminated soil can be restored by treatment with activated charcoal. You described areas in the U.S.A. of 30 acres or more with high concentrations of 2,4,5-T. Your laboratory experiments required 4 or 5 applications of 10^7 cells per gram of soil. Will treatment with bacteria of large areas involving thousands of tons of contaminated soil be practicable or even feasible?

A.M. CHAKRABARTY: Firstly, removing toxic chemicals by charcoal adsorption or various physical means from contaminated areas spanning several acres is impractical. Therefore a mechanism must be found that will allow ultimate removal of these compounds over a period of time.

I did not specifically mention it in my talk, but we have demonstrated that using lower concentrations of bacteria (10^4 to 10^6 cells per gram of soil), it is possible to remove more than 95% of 2,4,5-T provided one incubates the soil-bacteria mixtures for a longer period of time. Thus from a practical point of view, it is possible to use much less bacteria and achieve the same results. I should also add that the removal of 2,4,5,-T by Pseudomonas cepacia is just one example of the effectiveness of laboratory-developed cultures in pollution clean-up. I am sure future experiments, cloning of the 2,4,5-T degradative genes on a broad host range plasmid vector, and allowing the capability to be transferred among various soil Pseudomonads, for example, will greatly reduce the need for large-scale production of the original strain. Even if everything fails, considering that we do not have any alternative technology to deal with the problem, I think growing large quantities of the bacteria is within our technological capability and can be done from a practical point of view.

S. COHEN: Have you studied the genetic changes that are occurring during the 7 months of growth in the chemostat? Do you have an understanding of what genetic alterations have occurred during the evolution of a "previously non-existing" metabolic pathway?

A.M. CHAKRABARTY: This is of course one of the most interesting
questions we would have liked to answer. But unfortunately, for seven
months or so, there was only a very dilute population in the chemostat,
that one fine morning started to grow. So the only way we can study
the problem is to isolate the total DNA (both plasmid and chromosomal)
and determine the extent of homology of such DNA with the various
plasmid DNA sequences that we introduced during the selection
processes. If a specific plasmid gene was altered to help evolve a
specific gene in the 2,4,5-T degradative pathway, and if the
alteration is not extensive, then it might be possible to demonstrate
this evolutionary change by Southern blot hybridization. Assuming
that several such hybridizable sequences can be detected in the
plasmid DNA and the 2,4,5-T degradative genes which might have evolved
either on a plasmid (there is a plasmid in P. cepacia AC1100, although
we do not know its role in 2,4,5,-T degradation), on the chromosome, or
partly on the plasmid and partly on the chromosome, it might even be
possible to determine what genetic sequence, encoding a particular
enzyme, might have been altered to give rise to a new gene encoding a
new enzyme. Such studies are, however, for the future. At present we
have very little understanding about the evolutionary genetic changes
that might have taken place in the chemostat.

P. LAGGNER: The benefits of pseudomonas strain A1100 in partial soil
decontamination are great and obvious. However, there seems to be also
a risk involved: the availability of A1100 or future, even more
efficient strains, could reduce the threshold for the use of poisons
like 2,4,5-T in tactical military operations. This would far outweigh
the prima facie benefits.

A.M. CHAKRABARTY: I think the risk-benefit ratio will always guide the
development of strains like the 2,4,5-T degrading strain. To be sure,
the military would like to use various poisons for tactical warfare
(although biological or chemical warfare is banned under the Geneva
Convention), but unless they find a mechanism to remove the poisons
after their intended effect, they have no business in using such
materials. I am not too concerned about what problems the development
of poison-eating bacteria may cause to the military for making their
poisons non-functional.

P. LAGGNER: You showed that 2,4,5-T can act as a growth stimulator to
plants at very low doses.   Is there anything known about the
biochemical mechanisms of this effect?

A.M. CHAKRABARTY: A large amount of data are available in the
literature regarding the mode of action of 2,4,5-T on plants.

W. van den DAELE: Before genetically manipulated organisms can be
released into the environment, the safety of such release has to be
proven.   What would be the normal steps in such a proof?   In
particular, what would be appropriate safety measures, given the fact
that it is difficult to predict behavior in the environment from
laboratory experiments? Can one imagine schemes of safety research
that might be applied to meet public concern about possible risks?

A.M. CHAKRABARTY: You have to realize that bacteria such as the 2,4,5-T degrading strain are obtained from Nature. So they were there in the soil all the time. The only thing that has been introduced into them is a set of genes that allows them to utilize and completely metabolize some toxic chemicals like 2,4,5-T, a variety of chlorophenols, etc. Thus I do not see any potential danger in using a natural microorganism whose only change has been its ability to eat toxic chemicals. Of course, one will have to do a series of toxicological tests to demonstrate that the bacteria do not produce any harmful toxic effects on human beings, animals or even other microorganisms. We have demonstrated that once the 2,4,5-T is consumed, the strain dies off quickly because of its inability to compete with indegenous soil microflora. This eliminates any long-term harmful effects that use of such a strain may pose in the future.

DELIBERATE RELEASE OF MICROORGANISMS FOR INSECT CONTROL

P. Lüthy, H.M. Fischer and R. Hütter

Institute of Microbiology, Swiss Federal Institute
of Technology, 8092 Zürich, Switzerland

ABSTRACT

Viruses, bacteria and fungi are important regulators of
insect populations in the natural environment. Several
microbial preparations have passed registration and are
applied in crop protection and forestry. The safety record
of microbial insecticides is excellent. Up to date, not
a single case of development of resistance has been reported. The field concentration of pathogens following a
treatment is usually lower than after a natural outbreak
of the disease within the population of the target insect.

INTRODUCTION

Sufficient food for the steadily increasing world population can only be produced on a regular basis by the use of
plant protection agents. The crops need protection from
attacks by insects, fungi and weeds. Monoculture farming
and high productivity aggravate the pressure by pests,
especially by insects. However, the widespread use of
pesticides has a number of undesirable consequences such
as development of resistance in the target pest, disturbance of natural balances in the ecosystem caused by low
specificity of the pest control agents and finally problems with residues. The responsible authorities and industries are well aware of this fact, and over the last
years a clear trend away from hard broad spectrum chemicals towards more specific and biodegradable compounds can
be recognized.

Projects for the investigation of new insecticides include
the study of mechanisms by which nature controls insect
populations. Pathogenic microorganisms, insect growth regulators, pheromones and built-in resistances in plants
are the most promising research areas. The concepts of
nature cannot be adopted unaltered but have to be brought
in line with the requirements of modern crop management.
Efficiency and price as key factors must be competitive
with the classical chemicals while it is expected that the
so-called biological insecticides are without negative
side-effects.

CHARACTERIZATION OF VIRAL AND MICROBIAL
INSECT PATHOGENS

Microorganisms and viruses are the most important regulators of insect populations. They exert control either by
their continuous presence on an endemic level or by the
induction of epizootics.

Arthropod viruses are an abundant and diversified group.
To date, for 800 species of insects and mites more than
1200 host-virus relationships have been described (Martignoni and Iwai 1981). Insect viruses have as a rule a
narrow host range, not passing the bounderies of an order.
Many are so specific that they attack only one or a few
related species. The viruses are classified by the type of
nucleic acid (DNA, RNA) and the morphology of the virions.
A further criterion is the presence of an inclusion body
in which the virions are embedded either singly or in
bundles. The nuclear polyhedrosis (NPV) and the granulosis
viruses (GV) are the most important for insect control.
They are grouped together as baculoviruses and seem to
be associated only with invertebrates. A comprehensive
review on arthropod viruses is given by Krieg (1973).

Entomopathogenic fungi are omnipresent within insect populations and are able to reduce their number efficiently.
In contrast to viruses which have to be taken up orally,
fungi infect their hosts by penetration through the epi-
dermis. During infection the fungal pathogens often produce toxins (Roberts 1981). Entomopathogenic fungi are
encountered in the different taxonomic groups. Species
belonging to the genera Beauveria and Metarrhizium (fungi
imperfecti) possess a rather wide host range covering several orders. Others are more specific and limited in
their host range to a single family, genus or a few
species (e.g. Coelomomyces, Entomophthora). Outbreaks of
fungal epizootics can be observed regularly during the
vegetation period, for example within populations of
aphids where species of Entomophthora are the main pathogens.

Bacterial diseases among insect populations are rare and
the number of entomopathogenic bacteria is essentially
limited to a few species of spore formers with narrow
host ranges. Under field conditions bacteria are present
in an endemic state, as for example Bacillus popilliae,
a pathogen of the Japanese beetle (Popilliae japonica).
Epizootics may occur within confined areas such as in
insect populations attacking stored products. A typical
representative of this group is Bacillus thuringiensis.
The access of the bacteria to the hemocoel of the hosts
is frequently facilitated by preformed toxins (Lüthy 1975).

Finally, many insects are susceptible to diseases caused
by protozoans which predominantly produce chronic infections leading to a reduction of the life-span and fertility. For direct and fast insect control they are not

applicable with the exception of maybe <u>Nosema locustae</u>, a pathogen of grasshoppers (Henry and Oma 1981).

VIRAL AND MICROBIAL INSECTICIDES IN USE AND THEIR BEHAVIOR IN THE ECOSYSTEM

Some viruses, bacteria and fungi have passed all stages of process development and safety tests, and are commercially produced while others are in the process of registration or are tested on a large scale experimental level. The most important insect pathogens are listed in Table 1. Among the viruses emphasis is put on the inclusion body forming baculoviruses since they possess a number of required properties, such as sufficient stability, relative ease of production and high virulence. Up to date all virus material is produced in massrearings of host insects. No tissue culture systems have been developed to a stage where they could be used for commercial production. The main and most intensively investigated virus preparation is the NPV of <u>Heliothis</u> which comprises species of world-wide importance attacking at least 30 different food and fibre crops (Ignoffo and Couch 1981). The spectrum of the virus is limited to the genus <u>Heliothis</u>. The number of polyhedral inclusion bodies (PIB) needed for the control of <u>Heliothis</u> lies between 2.10^{11} and 2.10^{12}/ha, depending on the crop, the <u>Heliothis</u> species and the level of infestation. The activity of viruses does not persist for long on treated foliage. This was quantitatively investigated by Jaques (1975) with an NPV of the cabbage looper (<u>Trichoplusia ni</u>)(Fig.1). Loss of activity is due to

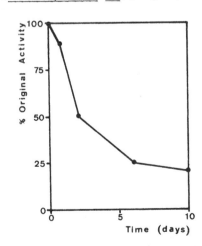

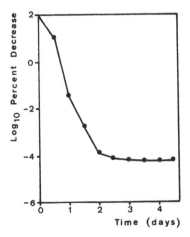

Figure 1. Decrease of activity following a NPV treatment against <u>Trichoplusia ni</u> on cabbage (after Jaques 1975).

Figure 2. Decrease of viable endospores of <u>B. thuringiensis</u> on a leaf surface (after Leong et al. 1980).

TABLE 1. The most important viral and microbial insecticides

Organism	Target insects	Main crops	Year of re-gistration	Remarks
Viruses:				
NPV (nuclear poly-hedrosis)	Heliothis spp.	cotton, corn, tobacco	1975	application rates average 10^{11}–10^{12} PIB/ha depending on crop, virulence of virus and insect density
NPV	Lymantria dispar	forest (oak)	1978	
NPV	Orgyia pseudotsugata	forest (Douglas fir)	1976	
GV (granulosis)	Laspeyresia pomonella	apple, pear	–	candidate for registration
Bacteria:				
Bacillus thuringien-sis (kurstaki and other varieties)	numerous lepidopterous pest insects	vegetable crops, tobacco, soja	1960	production $1,5.10^{6}$ kg/a application/ha: spores 10^{13}; δ-endotoxin 50g
Bacillus thuringien-sis var. israelensis	mosquito and black fly larvae		1981	control of vectors of tropical diseases
Bacillus sphaericus	mosquito larvae		–	experimental stage
Fungi:				
Beauveria bassiana	Coleoptera, Diptera	potatoes	–	in use in the USSR
Verticillium lecanii	aphids	chrysanthemum, vegetables	1981	applied in glasshouses
Hirsutella thompsonii	Phyllocoptruta oleivora	citrus	1981	application rates for fungi: 10^{12}–10^{13} spores/ha

irradiation by sunlight, high temperature and humidity.
On the other hand, Thomas (1975) found a substantial increase in PIB's in the soil five days after treatment
with the same virus. The soil is generally regarded as
reservoir for insect pathogens.

Bacillus thuringiensis (BT) is the only bacterium which
is produced on an industrial scale. The pathogenicity is
based on a preformed toxin, synthesized during the sporulation process and deposited in a crystalline form within
the sporangium. This metabolite with specific larvicidal
activity against lepidopterous or dipterous insects is a
protein and designated delta-endotoxin (Lüthy and Eber-
sold 1981). It is a potent poison destroying the gut epi-
thelium of susceptible larvae. One of the most interesting
properties of the delta-endotoxin is the fact that no resistance has developed during the many years of application.

The production of BT is easy and cheap since agricultural
waste products can be used for fermentations. Formulated
BT preparations contain viable bacterial spores along
with crystalline delta-endotoxin. BT products are used in
agriculture and to a lesser extend in forestry to reduce
the degree of defoliation by lepidoptereous insects.
An agriculturally important pest which is efficiently controlled by BT is the cabbage looper (T. ni), a polypha-
geous insect attacking various vegetable crops. The majority of the commercial preparations contains the BT variety kurstaki which produces a delta-endotoxin with a
high specific activity against the important target insects.

During a regular BT treatment about 5.10^{13} spores/ha are
disseminated along with the toxic protein crystals. On
the leaf surface the spores show low persistence (Leong
et al. 1980) and loose viability readily, mainly through
inactivation by sunlight (Fig.2). The life span of spores
reaching the ground during application is somewhat higher.
DeLucca et al. (1981) found in an extensive survey in the
US that BT spores were present at a higher rate in fields
treated previously. For example in a field treated two
years earlier, one colony out of a total of 141 sporeformers
was determined as B. thuringiensis. In an experimental
field, frequently sprayed with BT, 252 out of 1469 bacilli
were B. thuringiensis. Studies by Saleh et al. (1970) indicate, however, that B. thuringiensis does not become
part of the indigenous microflora.

A few years ago, a new BT variety, named israelensis, with
high and specific activity against mosquito and black fly
larvae was discovered by Goldberg and Margalit (1977).
Among the susceptible larvae we find important vectors of
human diseases, e.g. malaria and onchocerciasis (river
blindness). Commercial preparations are already available
and large scale treatments of infested waters are in progress, supported by funds of WHO. Although no systematic

investigations on the persistence of BT spores in water exist, our own data, summarized in Table 2 show that spores of the variety israelensis could be reisolated from all sites tested, even 18 months after the treatment. No israelensis spores were present previously in this area. The number of spores is, however, too small to give permanent control of mosquito larvae.

Since the delta-endotoxin is primarily responsible for the pathogenicity of B. thuringiensis, the presence of spores is not required in commercial products. We have developed in our laboratory asporogenic, delta-endotoxin producing mutants of the varieties kurstaki and israelensis (Cordier 1982). Even if BT spores are harmless, any microbial contamination, especially of water, is undesirable and should be kept at a minimum level. According to preliminary tests, our asporogenic mutants should fulfill the requirements for large scale production.

TABLE 2. Recovery of B. thuringiensis var. israelensis from water treated at a rate of 10'000 spores/ml

Water body	Volume (m^3)	Months after treatment	spores/ml recovered
Ditch	3	2	14
Pond	6	12	385
Small lake	400	18	19

Fungal insecticides are applied up to date only to a minor extent. In the USSR Beauveria bassiana is produced on a large scale for the control of the Colorado beetle (Leptinotarsa decemlineata). Recently, two fungal preparations have been registered, one in the UK for the control of aphids in glasshouses, and the other in the US against the citrus rust mite (Phyllocoptruta oleivora) on citrus. The use of fungal insecticides for the protection of glasshouse crops is promising since high humidity necessary for spore germination on the surface of the insect host is usually realized. In addition, the fungal spores are well protected under glass against UV-light by which they are otherwise quickly inactivate. Fungi are applied in the same order of magnitude as viruses and bacteria (10^{12} - 10^{13} spores/ha)

SAFETY CONSIDERATIONS

Absolute safety can never be guaranteed neither for a chemical nor for a biological insecticide since both have to interfere with living systems. The toxicological and ecological safety record of viral and microbial

insecticides is excellent (Burges 1981). Their harmless-
ness to non-target organisms has not only been proved in
stringent safety tests, but also during the long-term
use of these agents without a case of illness or death.

Following release, the insecticides become part of the
environment and their further behavior which has already
been discussed in the previous section, is the basis for
the evaluation of ecological safety. In contrast to xeno-
biotic chemicals, microorganisms in use for pest control
occur also naturally in concentrations which are often
higher than after deliberate release (Jaques 1981). According to Franz and Krieg (1982) the number of polyhedral
inclusion bodies after an outbreak of a virus epidemic
within an insect population can amount to 10^{20}/ha. Heimpel
et al. (1973) found 10^6PIB's/cm^2 of a T. ni NPV on untrea-
ted cabbage leaves from market shelves in Maryland.

Since baculoviruses and the bacterium B. thuringiensis
are the most widely used microbial insecticides, some
aspects of safety related to these organisms will be discussed in more detail:

The host specificity of viruses is fundamental for safety.
A cascade of mechanisms has to function to permit virus
multiplication, starting with a multistep uptake system
followed by production of virus specific protein and by
DNA replication. The virus has then to be assembled within
the cell and a mechanism for release must exist. Non-target organisms mostly lack a suitable uptake system or do
not permit cell internal virus replication. The specificity of baculoviruses was checked in vivo and in tissue
culture. Cell lines derived from mammals, birds, fish,
reptiles and amphibians did neither support virus multiplication (McIntosh 1975) nor were metabolic processes
affected (Arif and Dobos 1979). The few reports (e.g.
Hymeno et al. 1967) where interaction of baculoviruses
with non-host cells was obtained could not be reconfirmed.
Therefore the presently used virus preparations can be
considered as safe (see also Summers et al. 1975; Schmidt
and Winkelbrandt 1978; Krieg et al. 1980).

During virus multiplication interactions between the desired virus and the host DNA or other concomitantly present viruses may occur (Arber 1979). Hybrid viruses might
arise as a consequence. Recombinations are most likely to
occur in nature and are inevitable.Therefore virus prepa -
rations should be checked as part of the quality control
for the presence of foreign DNA.

Also for B. thuringiensis host specificity is fundamental
for safety. Spores and vegetative cells have proved harmless in infectivity tests with mammals. No multiplication
was observed and the cells were readily eliminated. No
toxic activity could be demonstrated for the native delta-
endotoxin outside the host insects. Allergenic reactions
have not been reported neither for the bacterium nor for
the delta-endotoxin (Burges 1981).

The genetic stability of B. thuringiensis has not yet been discussed in the literature with respect to safety. It has been shown in our laboratory that intra- and interspecific plasmid transfer in B. thuringiensis and with Bacillus subtilis can occur (H.M. Fischer, unpublished), but the rate of genetic transfer is low ($\leqslant 10^{-6}$). On the other hand some B. thuringiensis strains exhibit high resistance to a number of antibiotics (Martin 1979). The variety kurstaki, used in most commercial BT preparations, is resistant towards ampicillin, spectinomycin, polymyxin, sulfonilamide and thrimethoprim. The genetic location of the resistances (chromosomal or plasmids) is not known. With B. subtilis the transfer of antibiotic resistance markers has been shown to occur also in soil (Graham and Istock 1979). It is uncertain to which extent large scale application of BT preparations might influence the antibiotic resistance pattern. Some investigations in this direction seem desirable.

FUTURE DEVELOPMENTS

The trend to softer and more specific insecticides will continue or even be reinforced, resulting possibly in the development of more viral, bacterial and fungal products. Research institutions of universities and industries have started with the genetics of B. thuringiensis where rapid progress has been achieved. The delta-endotoxin genes have been cloned in E. coli (Schnepf and Whiteley 1981) and in B. subtilis (Klier et al. 1982) and part of the DNA sequence of the delta-endotoxin gene has been determined (Wong et al. 1983). Genetic engineering may lead to strains with higher toxin yields or enhanced specific activity, to hybrid delta-endotoxins with altered host spectra or to the production of the metabolite by vegetative cells.

ACKNOWLEDGMENTS

Our research with B. thuringiensis is supported by grants of the Swiss National Science Foundation and the WHO.

REFERENCES

Arber,W.(1979). Promotion and limitation of genetic exchange. Science 205, 361-365

Arif,B.M.and Dobos,P.(1979). Effect of a nuclear polyhedrosis virus of the spruce budworm on the metabolic processes of vertebrate cells. Forest Pest Management Inst., Sault Ste. Marie, Ont. Canada Report FPM-X-15, pp.16.

Burges,H.D.(1981).Safety testing and quality control of microbial pesticides, in Microbial Control of Pests and Plant Diseases 1970-1980 (Burges,H.D.ed) pp. 737-767, Academic Press, London and New York.

Cordier,J.L.(1982). Sporulationsmutanten von Bacillus thuringiensis. Diss. ETH Nr.7199, Swiss Federal Institute of Technology, Zürich, Switzerland.

DeLucca,A.J.,Simonson,J.G.and Larson,A.D.(1981). Bacillus thuringiensis distribution in soils of the United States. Can.J.Microbiol.27,865-870.

Franz,J.M.and Krieg,A.(1982). Biologische Schädlingsbekämpfung, pp.252, Paul Parey, Berlin and Hamburg.

Goldberg,L.J.and Margalit,J.(1977). A bacterial spore demonstrating rapid larvicidal activity against Anopheles sergentii, Uranotaenia unguiculata, Culex univittatus, Aedes aegypti and Culex pipiens. Mosquito News, 37,355-358.

Graham,J.B.and Istock,C.A.(1979). Gene exchange and natural selection cause Bacillus subtilis to evolve in soil culture. Science 204, 637-639.

Heimpel,A.M.,Thomas,E.D.,Adams,J.R.and Smith,L.J.(1973). The presence of nuclear polyhedrosis virus of Trichoplusia ni on cabbage from the market shelf. J. Environ.Entomol.2,72-75.

Henry,J.E.and Oma,E.A.(1981). Pest control by Nosema locustae, a pathogen of grasshoppers and crickets, in Microbial Control of Pests and Plant Diseases 1970-1980 (Burges,H.D.ed) pp.573-586, Academic Press New York and London.

Himeno,M.,Sakai,F.,Onodera,K.,Nakai,H.,Fukuda,T.and Kawade,Y.(1967). Formation of nuclear polyhedral bodies and nuclear polyhedrosis virus of silkworm in mammalian cells infected with viral DNA. Virology 33, 507-512.

Ignoffo,C.M.and Couch,T.L.(1981). The nucleopolyhedrosis virus of Heliothis species as a microbial insecticide, in Microbial Control of Pests and Plant Diseases 1970-1980 (Burges,H.D.ed)pp.329-362, Academic Press, London and New York.

Jaques,R.P.(1975). Persistence, accumulation, and denaturation of nuclear polyhedrosis and granulosis viruses, in Baculoviruses for Insect Pest Control: Safety Considerations (Summers,M.,Engler,R.,Falcon,L.A.and Vail,P.eds)pp.90-101,American Society for Microbiology, Washington.

Klier,A.,Fargette,F.,Ribier,J.and Rapoport,G.(1982). Cloning and expression of the crystal protein genes from Bacillus thuringiensis strain berliner 1715. EMBO J.1,791-799.

Krieg,A.(1973). Arthropodenviren. pp.328, Georg Thieme, Stuttgart.

Krieg,A.,Franz,J.M.,Gröner,A.and Huber,J.(1980). Safety of entomopathogenic viruses for control of insect pests. Eviron.Conserv.7,158-160.

Leong,K.L.H.,Cano,R.J.and Kubinski,A.M.(1980). Factors affecting Bacillus thuringiensis total field persistence. Environ.Entomol.9,593-599.

Lüthy,P.(1975). Zur bakteriologischen Schädlingsbekämpfung: Die entomopathogenen Bacillus-Arten, Bacillus thuringiensis und Bacillus popilliae. Vjschr. naturf.Ges.Zürich 120,81-163.

Lüthy,P.and Ebersold.H.R.(1981). The entomocidal toxins of Bacillus thuringiensis. Pharmac.Ther.13,257-283.

Martignoni,M.E.and Iwai,P.J.(1981). A catalogue of viral diseases of insects,mites and ticks, in Microbial Control of Pests and Plant Diseases 1970-1980 (Burges,H.D.ed) pp.897-911, Academic Press, London and New York.

Martin,P.A.W.(1979). A genetic system for Bacillus thuringiensis. pp.124, Diss.Ohio State Univ.(Order No. 8009309)

McIntosh,A.H.(1975). In vitro specificity and mechanisms of infection,in Baculoviruses for Insect Pest Control:Safety Considerations (Summers,M.,Engler,R. Falcon,L.A.and Vail,P.eds)pp.63-72, American Society for Microbiology, Washington.

Roberts,D.W.(1981). Toxins of entomopathogenic fungi, in Microbial Control of Pests and Plant Diseases 1970-1980 (Burges,H.D.ed) pp.441-464, Academic Press, London and New York.

Saleh,S.M.,Harris,R.F.and Allen,O.N.(1970). Fate of Bacillus thuringiensis in soil:effect of soil pH and organic amendment. Can.J.Microbiol.16,677-680.

Schmidt,L.and Winkelbrandt,E.(1978). Safety Aspects of Baculoviruses as Biological Insecticides,pp.301, Herausgeber:Bundesministerium für Forschung und Technologie,5300 Bonn,West-Germany.

Schnepf,H.E.and Whiteley,H.R.(1981). Cloning and expression of the Bacillus thuringiensis crystal protein gene in Escherichia coli. Proc.Natl.Acad.Sci.USA 78,2893-2897.

Summers,M.,Engler,R.,Falcon,L.A.and Vail,P.(1975). Baculoviruses for Insect Pest Control: Safety Considerations. pp.186, American Society for Microbiology.

Wong,H.C.,Schnepf,H.E.and Whiteley,H.R.(1983). Transcriptional and translational start sites for the Bacillus thuringiensis crystal protein gene. J.Biol. Chem.258,1960-1967.

DISCUSSION

P. STARLINGER: How much potential do you see for genetic engineering as opposed to conventional strain selection in B. thuringiensis?

P. LÜTHY: Genetic engineering opens possibilities for which there is no approach with conventional strain selections. For example only a part of the whole polypeptide (40%) represents the actual toxin. By elimination of the surplus DNA of the crystal protein, strains with increased production capacity for the toxin could be constructed. The production of delta-endotoxin is closely linked to the sporulation process. Strains excreting this metabolite into the medium during the vegetative growth phase, which is an interesting alternative, can only be obtained by genetic engineering.

L. OTTEN: When using insect viruses to control insects, should one not be concerned about possible effects on useful insects?

P. LÜTHY: It is one of the advantages of viral and microbial insecticides that beneficial insects are not harmed. They are able to continue their important function as predators or parasites. The nuclear polyhedrosis viruses which are used for insect control are highly specific, attacking only closely related species. The host range of a given nuclear polyhedrosis virus does not go beyond an insect family.

GENE EXPRESSION IN PLANTS

ISOLATION OF TRANSPOSABLE ELEMENTS IN MAIZE

P.Starlinger,U.Courage-Tebbe,   H.P.Döring,   W.B.-Frommer, K.Theres, E.Tillmann, E.Weck, W.Werr

Institut für Genetik, Universität zu Köln, 5000 Köln 41, FRG

ABSTRACT

In plants, particularly in maize, transposable elements are characterized by genetic tests, but no products of their genes are known. Methods to isolate such elements are described and illustrated mainly by work on traansposable element Ds inserted at the Shrunken locus in Zea mays.

The genetic phenomenon of transposition was discovered by McClintock in Zea mays (McClintock, 1951, 1956, 1965). This work was carried out at a time, when the biochemical analysis of these elements was still impossible. The biochemical study of transposition started with bacterial IS-elements (Jordan, Saedler and Starlinger, 1968; Shapiro, 1969). It was followed by the work on bacterial transposons and later by the biochemical study of transposition in several eukaryotes, particularly yeast, Drosophila melanogaster (reviews:Movable Genetic Elements, Cold Spring Harbor Symp. 1980).

Is it interesting to extend these studies to the biochemistry of plant transposable elements? We want to quote four reasons for this:

1. The amount of genetic and physiological knowledge on maize transposable elements is still more extensive than for any other transposable elements in other organisms.

2. The original observations of transposable elements were made in maize lines that had been known for many generations and had not shown any sign of transposition. The first observation of transposition was correlated in several cases with the occurrence of chromosome breaks. These chromosome breaks could be introduced either by the breakage-fusionbridge cycle (McClintock, 1951) or by ionizing radiation (Peterson, 1953). This is a most important observation. If it could be extended to other species, including higher animals and man, it could mean, that low-level radiation might be capable to start a continued system of mutation. This would be most important for our estimation of the effects of ionizing radiation.

3. The creation of genetic variety is important for the breeder.
Transposition mutagenesis has become a most useful tool in
bacteria. A more detailed knowledge of transposition in plants
might enable us to use transposition for the creation of new
genetic variation.

4. DNA injected into cells and particularly nuclei does not
automatically become part of the chromosome. This will in most
cases be a prerequisite for their inheritance, however. Genes
might be inserted into transposons which could then serve as
vectors for injected DNA into chromosomes. In Drosophila, this
method has been used successfully (Rubin and Spradling, 1982;
Spradling and Rubin, 1982) and is described elsewhere in this
volume. The isolation of plant transposable elements could make
this method amenable to plants.

With these aims in mind, we and others have embarked upon a
biochemical study of plant transposable elements.

The usual method to isolate genes of eukaryotes from a rather
complex DNA starts from the study of the gene product, usually
a protein. Its messenger RNA is identified, transcribed into
cDNA and cloned in the plasmid. The plasmid, containing the
cDNA insert, can then be used as a probe to screen
large genomic libraries for the appropriate genomic clones.

This method is not available for plant transposable elements, because their gene products are not known. We
therefore chose an indirect way. We tried to identify a
gene, which on the one hand encoded an abundant protein
and of which, on the other hand, mutants were available
that were caused by the insertion of a transposable element. It could then be hoped to isolate the gene both from
the wild type and from the mutant. DNA present in the
latter, but not in the former, would be a candidate for
the transposable element and could be used for further
studies.

Of the large collection of mutants that were isolated by
McClintock, Peterson and others, the gene Shrunken (Sh)
seemed most promising. Shrunken encodes the enzyme endosperm sucrose synthase, a key enzyme of starch biosynthesis
(Chourey and Nelson, 1976). Several Ds-induced mutants
were described by McClintock (1952, 1953). We considered
this to be an advantage offsetting the disadvantage that
the genetics of these mutants is more complicated than
that of some other Ds-induced mutations (McClintock, 1952,
1953).
After some difficulties, several laboratories, including
ours, managed to isolate cDNA clones (Burr and Burr, 1980,
1981; Geiser et al., 1980; Chaleff et al., 1981; Wöstemeyer
et al., 1981; Döring et al., 1981; Chaleff et al., 1983).
These could then be used for the isolation of wild type

clones (Burr and Burr, 1982; Geiser et al., 1982) as well
as clones from mutants (Geiser et al., 1982). Cloned DNA
could be characterized by restriction mapping and by DNA
sequence analysis. In addition, this DNA could serve as a
probe for blotting experiments with genomic DNA. Using
these methods, we arrived at the following conclusions up
to now:

Our mutant clone obtained from sh-m5933, one of the shrunken mutants, consists of two segments. One of these is
homologous to the corresponding segment of our wild type
clone. The other one has a length of 15 kb and extends
from a junction point to the end of the insert in the
phage. It has no homology to the Sh wild type clone and
thus will be called "foreign" DNA henceforth.

It is important to determine whether part or all of the
foreign DNA is Ds. At present, we believe that the first 4
kb adjacent to the junction point with wild type DNA are
Ds, and that the rest of the "foreign" DNA does not belong
to Ds, but indicates a complicated structure of the mutation in sh-m5933.
The 4 kb sequence of "foreign" DNA adjacent to the breakpoint with wild type DNA consists of two identical sequences of 2 kb each. One of these is inserted in inverted
orientation in the second one, yielding a structure that
can be represented as D (s'D') s. Each of these 2 kb
sequences is terminated by an inverted repeat of 11 bp
length. Adjacent to both termini of the central 2 kb
sequence, a direct repeat of 8 bp is present. These 8 bp
are present within the central copy only once. The 2 kb
sequence with 11 bp inverted repeat at the termini thus
behaves like many other transposons, leading to a short
duplication of host DNA upon insertion. (H.P.Döring, E.Till-
mann, P.Starlinger, unpublished experiments).

A mutation caused by Ds in Adh1 was isolated by Ostermann
and Schwartz (1981) and has been cloned and sequenced by
J.Peacock and collaborators (personal communication). The
insert in this mutation is 402 bp long and terminates in a
11 bp repeat identical to the 11 bp inverted repeat found
at the termini of the transposon-like structure in sh-
m5933. 8 bp of Adh1 DNA adjacent to the point of insertion
are duplicated in the mutant. This supports the assumption
that the 4 kb sequence at the junction of foreign and
shrunken DNA in sh-m5933 is Ds and raises the question of
the nature of the rest of the foreign DNA.

Another Ds-induced mutation in gene sh is shm6233. Preliminary genomic blotting analysis (U.Courage-Tebbe, H.P.Döring
and P.Starlinger, unpublished experiments) indicates that
this mutation carries a 4 kb insert. The order of restriction sites indicates a similar but not completely identical
structure as the D(s'D')s found at the breakpoint in

sh-m5933. This mutation may thus be simply an insertion of a double Ds deviating from the double Ds in sh-m5933 by an internal inversion. E.Weck has isolated a clone carrying part of this insert at the 3'-terminal end of shrunken including the junction. Sequence analysis shows the 11 bp adjacent to the breakpoint to be identical and those found in sh-m5933 and the Ostermann-Schwartz mutant in Adh1.

These results indicate strongly that all of these mutants are caused by the insertion of Ds, a transposon terminating in an 11 bp inverted repeat and causing 8 bp duplications at the site of insertion. The different copies of Ds are not of identical size, however. They can vary from 2 kb, the unit length in sh-m5933 to 402 bp in the Ostermann-Schwartz mutant. Transposons with identical termini and variable length may be formed from a larger original transposon by internal deletions. A similar situation has been observed in the case of the P-element in Drosophila melanogaster (Spradling and Rubin, 1982). Here, it is speculated that a master P-element gives rise to partial deletions. Exactly the same situation may be underlying the two-element transposons of McClintock, which are often observed after being initiated as a one-element system. Among others this is true for Ac and Ds. We thus anticipate that Ac will be similar to Ds but longer than or identical in length to the largest Ds isolated so far. The various copies of Ds may thus be complementable mutants of Ac.

Blotting experiments using Ds-DNA as a probe and genomic DNA of maize and teosinte   show up to 40 hybridizing bands after digestion with various restriction enzymes (Geiser et al., 1982; K.Theres, unpublished experiments). This indicates that Ds is present in several copies. The exact number is not determined but may be of a similar order of magnitude in the hybridizing bands. Their presence in teosinte   indicates that Ds is evolutionarily old. The origin of Ds and Ac in maize lines apparently devoid of this transposable element system may thus indeed be an activation of a previously silent copy of Ac.

Ds-induced mutations can be more complicated than a simple insertion. The formation of double Ds structures was already mentioned. sh-m5933 is more complicated than this. The clone isolated from the mutant shows more than 10 kb of "foreign" DNA adjacent to the double Ds structure. Part of this has been used as a probe in genomic blotting experiments and has given a single band only after digestion with certain restriction enzymes. Thus, unique DNA seems to be associated with Ds in the insert carried in Sh in mutant sh-m5933 (K.Theres, unpublished experiments).

Together with N.Fedoroff, we determined the structure of
sh-m5933 in more detail and found that the mutant carries
a 30 kb insert within Sh (Courage-Tebbe, Döring, Fedoroff
and Starlinger, manuscript in preparation). The 15.5 kb of
"foreign" DNA cloned previously are part of this insert.
Several restriction sites on the other side of the insert
have been determined by blotting analysis. They resemble
closely those found in Ds at the breakpoint adjacent to
the 3'-end of gene sh, indicating the possible presence of
another Ds copy at the other breakpoint of the insertion.

sh-m5933 is even more complicated: Part of the insert, the
5'-end of gene Sh adjacent to this insert and at least 20
kb of DNA adjacent to it are duplicated in the mutant.
N.Fedoroff has isolated revertants to Sh. In these revertants, the 30 kb insert in gene Sh is lost, restoring the
restriction pattern of the wild type. The duplication
including the 5-end of Sh is retained, however. The fact
that these revertants still give rise to chromosome breaks
at the locus of Sh could be explained if we assume that
the duplicated copy is located near the gene Sh and that
Ds is indeed located at the breakpoint between the "foreign" DNA at the 5'-end of gene Sh in this duplicated
copy (Courage-Tebbe, Döring, Fedoroff and Starlinger, manuscript in preparation).

In summary, Ds has been cloned from a variety of sources,
has been sequenced and seems to be capable of transposition
provided that its termini are present. Much of its interi-
or, however, seems to be unnecessary for transposition.
Thus, Ds seems to be a suitable material for the construction of artificial transposons consisting of the termini
of Ds and other DNA located in between. Such experiments
are presently being tried in several laboratories including
ours.

The availability of Ds as a hybridization probe might be a
means to isolate other genes, provided Ds-induced mutations
are at hand. This obvious approach of using Ds as a handle
to isolate mutant DNA will be made difficult, however, by
the presence of many copies of Ds in each genomic complement of maize DNA. The number of bands seen in blotting
experiments with Ds-DNA as a probe is not prohibitively
large, however.

Two approaches are offering themselves. If a particular
gene product is known already and if this can be identified
in cell-free protein synthesis, many clones containing Ds
may be isolated from a genomic library and tested separately in hybrid-released translation experiments.

If no gene product is available, the situation is more difficult. The comparison of the genomic blots of two lines carrying the Ds-induced allele of the gene in question and its wild type allele, respectively, could reveal which band corresponds to the mutated gene. As different maize lines yield slightly variable patterns, however, this approach is useful only, if the two maize lines are isogenic. Such lines will not usually be available.

We have devised an experiment that can get around the necessity of isogenic strains. We construct an F1 heterozygous for the mutant allele of the gene carrying the Ds-induced mutation that we want to isolate. From this F1, F2 is obtained by selfing, and several kernels each carrying the mutant or the wild type allele are planted. In most cases, it will be necessary to determine which of the plants are dominant homozygous for the wild type by selfing or back-crossing these F2-plants. A mixed DNA preparation is then obtained, using material from approximately 10 plants each of either genotype. This mixture is complex enough to average out most of the differences in Ds content and thus is a substitute for isogenic strains. The only difference between the two DNA preparations obtained by appropriate selection of the plants is the difference between the two alleles in question. K.Theres in our laboratory has shown in several instances that these mixed DNA preparations yield restriction patterns identical with the exception of one band only. It is presently being tried to clone these bands in which the two DNA preparations differ and thus to obtain the appropriate DNA fragment.

In summary, we are now in the position, where transposable element Ds is present as DNA in pure form. We hope to use it for the in vitro manipulation of this element in order to use it as a gene vector. In addition, we hope to use it as a handle to isolate yet unknown genes that are marked by the presence of the Ds element.

We appreciate the collaboration with N.Fedoroff and M.Freeling. We thank B.McClintock not only for strains, but for continued advice on the genetic experiments.
This work was supported by the Landesamt für Forschung des Landes Nordrhein-Westfalen, by the Commission of the European Community, and by Deutsche Forschungsgemeinschaft through Sonderforschungsbereich 74.

REFERENCES

Burr, B. and Burr, F. (1980). Detection of changes in maize DNA at the Shrunken locus due to the intervention of Ds elements. Cold Spring Harbor Symp. Quant. Biol. 45, 463-465

Burr, B. and Burr, F. (1981). Controlling-element events at the Shrunken locus in maize. Genetics 98, 143-156

Burr, B. and Burr, F. (1982). Ds controlling elements of maize at the Shrunken locus are dissimilar insertions. Cell 29, 977-986

Chaleff, D.,Mauvais, J., McCormick, S., Shure,M, Wessler, S., Fedoroff, N. (1981). Controlling elements in maize. Carnegie Inst. Wash. Year Book 80, 158-174

Chourey, P. and Nelson, O. (1976). The enzymatic deficiency conditioned by the shrunken-1 mutations in maize. Biochem. Gen. 14, 1041-1055

Döring, H.P., Geiser, M., Starlinger, P. (1981). Transposable element Ds at the shrunken locus in Zea mays. Mol. Gen. Genet. 184, 377-380

Fedoroff, N., Mauvais, J., Chaleff, D. (1983). Molecular studies at the Shrunken locus in maize caused by the controlling element Ds. J. Mol. Appl. Gen., in press

Geiser, M., Döring, H.P., Wöstemeyer, J., Behrens, U., Tillmann E., Starlinger, P. (1980). A cDNA clone from Zea mays endosperm sucrose synthase mRNA. Nucl. Acids Res. 8, 6175-6188

Geiser, M., Weck, E., Döring, H.P., Werr, W., Courage-Tebbe, U., Tillmann, E., Starlinger, P. (1982). Genomic clones of a wild-type allele and a transposable element-induced mutant allele of the sucrose synthase gene of Zea mays L. The EMBO J. 1, 1455-1460

Jordan, E., Saedler, H., Starlinger P. (1968). O° and strong polar mutations in the transferase gal operon are insertions. Mol. Gen. Genet. 102, 353-363

Marx, J.L. (1983) A transposable element of maize emerges. Science 219, 829

McClintock, B. (1951). Chromosome organization and genetic expression. Cold Spring Harbor Symp. Quant. Biol. 16, 13-47

McClintock, B. (1952). Mutable loci in maize. Carnegie Inst. of
    Wash. Year Book 51, 212-219

McClintock, B. (1953). Mutations in maize. Carnegie Inst. of
    Wash. Year Book 52, 227-237

McClintock, B. (1956). Controlling elements and the gene. Cold
    Spring Harbor Symp. Quant. Biol. 21, 197-216

McClintock, B. (1965). The control of gene action in maize.
    Carnegie Brookhaven Symp. Biol. 18, 162-184

Peterson, P. (1953). A mutable pale green locus in maize.
    Genetics 38, 682-

Rubin, GM. and Spradling, AC. (1982). Genetic transformation of
    Drosophila with transposable element vectors. Science
    218, 348-353

Spradling, AC. and Rubin, GM (1982). Transposition of cloned
    P-elements into Drosophila germ line chromosomes. Scien-
    ce 218, 341-348

Shapiro, JA (1969). Mutations caused by the insertion of genetic
    material into the galactose operon of Escherichia coli.
    J. Mol. Biol. 40, 93-105

Wöstemeyer, J., Behrens, U., Merckelbach, A., Müller, M., Star-
    linger, P. (1981). Translation of Zea mays endosperm
    sucrose synthase mRNA in vitro. Eur. J. Biochem. 114,
    39-44

Reviews: Movable Genetic Elements. Cold Spring Harbor Symp.
          Quant.Biol. XLV (1981)

DISCUSSION

H. SAEDLER: Your definition of "revertant" is genotypically, but the
phenotype is still that of a revertant. Can you elaborate on this?

P. STARLINGER: The revertants have a non-shrunken phenotype, a normal
kernel, and sucrose synthase. Thus, with respect to the shrunken gene,
they are true revertants. There is, however, a copy of Ds present on
the chromosone 9, near gene Sh. Its presence is revealed by somatic
chromosome breaks in the presence of Ac. We now believe that this Ds
copy is part of the duplicated region retained in the revertants.

E. coli - A. tumefaciens SHUTTLE VECTORS

G.E. Riedel and D.A. Austen

Genetics Institute, 225 Longwood Avenue,
Boston, MA   02115 U.S.A.

ABSTRACT

E. coli - A. tumefaciens shuttle vectors are described which allow
the transfer of cloned sequences into the T-DNA of the Ti plasmid.
Subsequently, the sequences transferred into T-DNA can be transferred
into plant cells through the action of A. tumefaciens. These vectors
have been used to transfer the gene encoding nopaline synthase into
plant cells.

INTRODUCTION

A. tumefaciens is a gram-negative bacterium which infects plants and
causes crown gall disease (for a recent review, see Ream and Gordon,
1982). In the course of infection, A. tumefaciens transfers a
portion of its genome to the nuclear genome of each infected plant
cell. The DNA transferred by A. tumefaciens is called T-DNA. This
DNA is a 14-20 kilobase pair (kbp) fragment located on the large
resident Ti (tumor inducing) plasmid of A. tumefaciens. The T-DNA is
transcribed and its mRNA is translated within plant cells, and it
confers several phenotypes (for examples, see Garfinkel et al, 1981).
Furthermore, "extra" DNA can be inserted into the T-DNA and is
co-transferred into plant cell genomes as part of the T-DNA
(Hernalsteens et al, 1981). It is this last property of the T-DNA,
namely, its ability to co-transfer "foreign" DNA sequences inserted
within it into plant genomes, that makes A. tumefaciens such a useful
tool for molecular genetic studies in higher plant cells.

The use of the T-DNA of A. tumefaciens as a vector has been
restricted because the Ti plasmid (in which the T-DNA is located) is
very large, has multiple restriction enzyme recognition sites, and
does not replicate in E. coli. Several groups have devised ingenious
strategies to overcome these difficulties (for example, see Leemans
et al (1981) and Ruvkun and Ausubel, 1981). We describe below a
vector that minimizes the manipulations necessary to introduce cloned
DNA fragments into the T-DNA of A. tumefaciens, and subsequently into
plants.

RESULTS

E. coli cloning vectors can be introduced into the T-DNA of A.
tumefaciens through plasmid mobilization and homologous
recombination. Ditta et al(1980) have described a triparental
bacterial conjugation system which can be used to mobilize the broad
GM-4

host range, self-replicating plasmid pRK290 into many different gram-negative bacteria. The system (see Figure 1) consists of E. coli strains carrying pRK290 and pRK2013, respectively. pRK290 is capable of self-replication in many gram-negative species but lacks self- transfer functions. The second plasmid, pRK2013, is capable of self-replication in only a few gram-negative species, but carries self-transfer and plasmid mobilizing functions. When E. coli cells carrying pRK2013 or pRK290 are mixed together with A. tumefaciens strain A6 in a bacterial conjugation experiment, transconjugant A6 individuals carrying pRK290 arise at a frequency of 0.01 per recipient A6 cell. It is likely that pRK2013 first transfers itself into the E. coli cell carrying pRK290, then transfers itself, and pRK290, into the A. tumefaciens recipient. Once within A6, pRK2013 is not replicated and is lost by dilution during bacterial cell division. The plasmid pRK290 is replicated and stably maintained within A. tumefaciens under appropriate selection (tetracycline, 20 µg/ml).

The observations of Warren et al (1979) showed that ColE1 derivatives can be mobilized by supplying appropriate functions in trans. We therefore replaced pRK290 with a variety of E. coli cloning vectors (see Figure 2). These vectors are not replicated as independent plasmids in A. tumefaciens, but they do carry a fragment of T-DNA which provides homology for reciprocal recombination. When these plasmids were used in triparental bacterial conjugation experiments, they were mobilized into A. tumefaciens by pRK2013 and recombined with the T-DNA with a combined frequency of 10^{-6} to 10^{-4} per A. tumefaciens recipient cell. Southern blot analysis (data not shown) of several transconjugants from each different vector conjugation experiment showed in all cases that the transconjugant A. tumefaciens strains carried cointegrate (pTi::E. coli vector) plasmids. These cointegrate plasmids appeared to have arisen by a single recombination event within the T-DNA carried by the E. coli vector and its homologous region in the Ti plasmid.

A specific example of these observations is presented in Figures 3 and 4. The plasmid pGR413 is a derivative of the E. coli cloning vector pBR322; it contains the gene encoding nopaline synthase inserted in its Hin dIII site and a fragment of T-DNA (homologous to T-DNA in strain A6) inserted in its Eco RI site. Figure 4 diagrams the steps we believe occurred during the conjugation experiment which introduced pGR413 into A. tumefaciens strain A6:

1. pRK2013 transferred itself into the E. coli strain carrying pGR413.

2. pRK2013 transferred itself and mobilized pGR413 into A. tumefaciens strain A6.

3. pRK2013 was lost from A6 during bacterial growth and pGR413 recombined with T-DNA at its region of homology to form a pGR413::Ti cointegrate plasmid.

(E. coli cloning vector):: Ti cointegrate plasmids transfer the E.coli vector into plant cells. The plasmid pGR413 was constructed in order to assay whether cointegrate Ti plasmids formed by this procedure would transfer "foreign" DNA into plant cells. The plasmid pGR413 contains a Hin dIII fragment that carries an intact nopaline synthase (nos) gene, derived from A. tumefaciens strain C58. The nos gene is constitutively expressed in transformed plant cells and is not contained in strain A6, nor in the genome of an untransformed plant cell. Thus, production of nopaline by transformed plant tissue would be good preliminary evidence that foreign DNA was transferred into and expressed by plant cells.

The transconjugant strain ATR29 (see figure 4) carrying a pTi::pGR413 cointegrate plasmid was used to incite tumors in axenically grown Nicotiana tabacum cv. Xanthi plants. After 14 days tumors were excised, transferred to MS media (Murashige and Skooge, 1962) containing carbenicillin [500 μg/ml], incubated in the dark at 28°C for 2 passages of 10 days each, then transferred to MS media. The tissue was assayed for nopaline essentially as described by Otten and Schilperoort, 1978. Figure 5 shows that tumor tissues induced by ATR29 produce nopaline. This result constitutes strong preliminary evidence that "cointegrate" Ti plasmids can transfer "foreign" DNA into plant genomes and that this "foreign" DNA can be expressed within those cells.

E. coli vectors can undergo double recombination events with the T-DNA within A. tumefaciens. We were interested in determing whether we could use E. coli cloning vectors to specifically insert "foreign" DNA within the T-DNA, as well as to form site specific cointegrates. To facilitate these experiments we constructed the plasmid pDA10 (see figure 6). pDA10 consists of a) pBR322, b) a 3.9 kbp fragment of T-DNA from pTiA6, and c) a 1.1 kbp fragment from Tn5 which encodes kanamycin resistance. The plasmid is constructed in such a way that pBR322 has been inserted into the T-DNA fragment.

When pDA10 is transferred by plasmid mobilization into A. tumefaciens strain A6, stable transconjugants arise at a frequency of 10^{-7} to 10^{-6} per recipient A6 cell. These transconjugants can be divided into two groups: those which carry a pTi::pDA10 cointegrate plasmid and those in which pBR322 appears to be inserted into the T-DNA of pTiA6 at precisely the location it is inserted within pDA10. This second class of transconjugants is most simply explained by assuming that a double recombination event occurred between pDA10 and pTiA6. A surprising feature of these "double recombinants" is that they arise more frequently than do the cointegrate plasmids.

We are presently investigating the properties of the two classes of transconjugants containing pDA10 in both A. tumefaciens and higher plant cells.

DISCUSSION

The shuttle vectors described in this report and the method of introducing them into T-DNA of A. tumefaciens are of considerable use in molecular genetic studies of higher plant cells. The procedure

consists of one step, occurs with a reasonably high frequency and
requires inexpensive bacteriological media. Transconjugants can be
picked immediately from the conjugation selection media and used to
induce tumors in whole plants or to infect protoplasts. Detailed
experimental protocols and materials are available upon request. One
possibility raised by the high frequency transfer of some
ColEl-derived vectors is the construction of a library of plant genes
in a shuttle vector in E. coli, its transfer into A. tumefaciens by
bacterial conjugation, and the subsequent use of the A. tumefaciens
population to infect protoplasts with biochemical genetic lesions. We
are conducting experiments to improve the efficiency of transfer of
shuttle vector DNA into T-DNA with the hope of eventually being able
to clone plant genes on the basis of gene complementation within plant
cells.

ACKNOWLEDGMENTS

This work was initiated at the CSIRO Division of Plant Industry. where
G.E.R was supported by the CSIRO and the Helen Hay Whitney Foundation.
G.E.R. also thanks A. Kerr and M. Tate for their extensive help in
teaching him opine assays and producing the data in figure 5.

During the course of this work we learned that Luca Comai and his
colleagues have independently developed similar vectors and
procedures. Their results closely correlate with ours.

REFERENCES

Ditta, G., Stanfield, S., Corbin, D. and Helinski D. (1980). Broad
        host range DNA cloning system for gram-negative bacteria:
        construction of a gene bank of Rhizobium melioti. Proc.
        Natl. Acad. Sci. U.S.A. 77, pp. 7347-7351.

Garfinkel, D., Simpson, R., Ream, L., White, F., Gordon, M. and
        Nester, E. (1981). Genetic analysis of crown gall: fine
        structure map of the T-DNA by site-directed mutagenesis.
        Cell 27, pp. 143-153.

Hernalsteens, J., Van Vliet, F., De Beuckeleer, M., Depicker, A.,
        Engler, G., Lemmers, M., Holsters, M., Van Montagu, M. and
        Scheldl, J. (1980). The Agrobacterium tumefaciens Ti
        plasmid as a host vector system for introducing foreign DNA
        in plant cells. Nature 287, pp. 654-656.

Leemans, J., Shaw, C., Deblaere, R., DeGreve, H., Hernalsteens, J.,
        Maes, M., Van Montagu, M. and Schell, J. (1981).
        Intermediate vectors for site specific mutagenesis of
        Agrobacterium Ti plasmids and for transfer of genes to plant
        cells. J. Mol. App. Genet. 1, pp. 149-164.

Murashige, T. and Skooge, F. (1962). A revised medium for rapid
        growth and bio-assays with tobacco tissue cultures. Physiol
        Plant. 15, pp. 473-497.

Otten, L. and Schileproort, R. (1978). A rapid microscale method for the detection of lysopine and nopaline dehydrogenase activities Biochim. Biophys. Acta 527, pp. 497-500.

Ream, L. and Gordon, M. (1982). Crown gall disease and prospects for genetic manipulation of plants. Science 281, pp. 854-859.

Ruvkun, G. and Ausubel, F. (1981). A general method for site-directed mutagenesis in prokaryotes. Nature 289, pp. 85-88.

Warren, JG., Saul, M. and Sheratt, D. (1979). ColE1 plasmid mobility: essential and conditional functions. Molec. Gen. Genet. 170, pp. 103-107.

G.E. Riedel and D.A. Austen

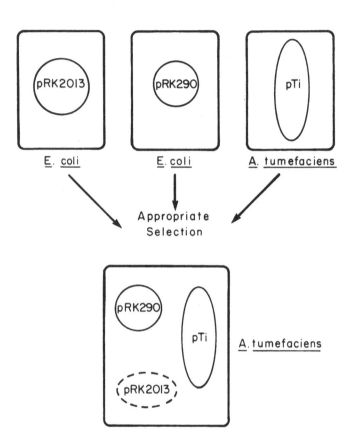

Figure 1. Mobilization of pRK290 into <u>A</u>. <u>tumefaciens</u> strain A6. The figure is explained in the <u>text</u>.

E.coli HYBRID PLASMIDS MOBILIZED
INTO A.tumefaciens A6

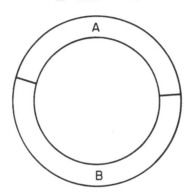

A = Eco RI Fragment ex - pTiACh5 Containing Tn5
Insertion

B = Col EI , pBR325 , Col EI::Tn7, pSF2I24 ,
pBR322 + nos

Figure 2. E. coli plasmids mobilized into A. tumefaciens
strain A6.

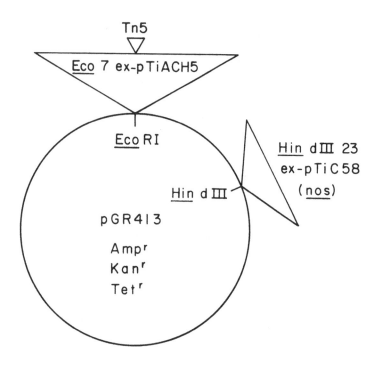

Figure 3. Physical map pGR413

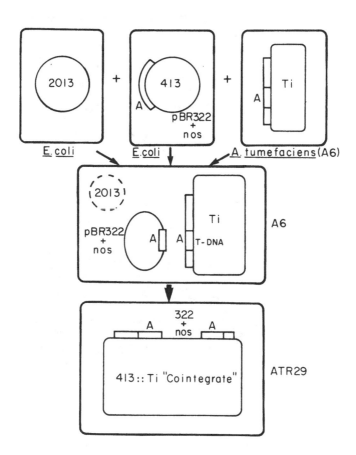

Figure 4. Mobilization of pGR413 into A. tumefaciens
strain A6. The figure is explained in the text.

　　　　　　　　G.E. Riedel and D.A. Austen

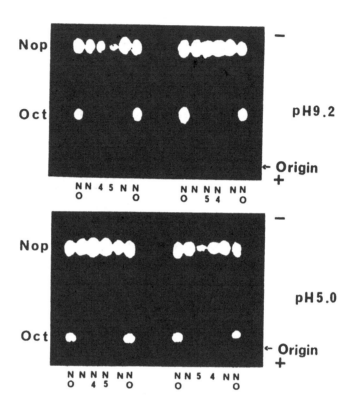

Figure 5. Nopaline assays of tumor tissues incited by
ATR29 (see text). Tissue extracts were
prepared, electrophoresed, and stained as
described by Otten & Schilperoort (1978).
Material from a spot with the relative mobility
of nopaline was eluted and the material was
re-tested in several buffer systems. The
numbers 4 and 5 refer to tumor tissue samples
incited by ATR29. O; octopine; N; nopaline.

E. coli - A. tumefaciens SHUTTLE VECTOR

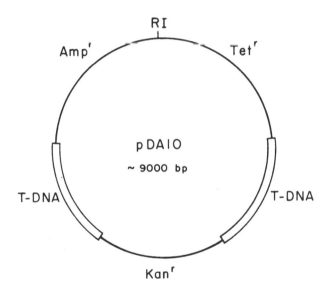

Figure 6. Physical map of pDA10

DISCUSSION

P. LUTHY: The gene of the insecticidal delta-endotoxin of <u>Bacillus</u> <u>thuringiensis</u> has been cloned in <u>E. coli</u>.   What are the chances of using the Tl plasmid as a vector to clone the delta-endotoxin gene in plants which are attacked by lepidopterous insects?   If the gene is expressed in a suitable form we would have plants carrying a non-toxic and very specific insecticide.

R. RIEDEL: The chances of doing this experiment are very good, and I am confident that several laboratories are pursuing this goal.

Ti PLASMIDS AS EXPERIMENTAL GENE VECTORS FOR PLANTS

J. Schell[1], M. Van Montagu[2], M. Holsters, P. Zambryski,
H. Joos, D. Inze, L. Herrera-Estrella, A. Depicker,
M. De Block, A. Caplan, P. Dhaese and E. Van Haute

Laboratorium voor Genetica,
Rijksuniversiteit Gent,
Gent, Belgium

J.P. Hernalsteens, H. De Greve, J. Leemans and R. Deblaere

Laboratorium voor Genetische Virologie,
Vrije Universiteit Brussel,
Sint-Genesius-Rode, Belgium

L. Willmitzer, J. Schroder and L. Otten

Max-Planck-Institut fur Zuchtungsforschung,
Koln, Federal Republic of Germany

Also affiliated to [1] the Max-Planck-Institut fur Zuchtungsforschung,
Koln (FRG); [2] to the Laboratorium voor Genetisch Virologie, Vrije
Universiteit Brussel (Belgium.

INTRODUCTION

The formation of so-called "crown gall" tumors on dicotyledonous
plants is the direct result of the introduction into the nuclear genome
of plant cells of a set of genes that regulates cell and organ
development. In other words, in Nature a mechanism exists that not
only efficiently introduces foreign genes into the plant nucleus, but
contains a set of genes which regulate plant cell development and
differentiation. As a result of this gene transfer, crown gall cells,
unlike untransformed plant tissues, can be cultured under axenic
conditions on synthetic media in the absence of growth hormones, i.e.
cytokinins and auxins.

This process is carried out by the so-called Ti plasmids of Agrobacterium. Most strains of the Agrobacterium genus, both pathogenic and nonpathogenic, contain one or more large plasmids (Zaenen et al., 1974; Currier and Nester, 1976; Merlo and Nester, 1977), many of which have remained uncharacterized. The different types of Ti plasmids which are responsible for the pathogenic properties of Agrobacterium all have a molecular weight in the range of 120 to 160 x 10⁶ D. A subgroup of Ti plasmids inducing hairy root tumors are often referred to as Ri plasmids. The tumor-inducing plasmids are most easily identified by their transfer to nonvirulent strains of bacteria (Van Larebeke et al., 1974; Watson et al., 1975). The transfer of virulence is always correlated with the transfer of a Ti plasmid (Kerr, 1969, 1971; Bomhoff et al., 1976; Hooykaas et al., 1977; Van Larebeke et al., 1977; Genetello et al., 1977; Kerr et al., 1977; Holsters et al., 1978). The tumor cells also produce low molecular weight compounds, called opines, not found in untransformed plant tissues. The type of opine produced defines crown galls as octopine-, nopaline-, or agropine-type tumors (Guyon et al., 1980). Transfer experiments have demonstrated that Ti plasmids are responsible for most of the typical properties of agrobacteria: (i) crown gall tumor induction; (ii) specificity of opine synthesis in transformed plant cells; (iii)catabolism of specific opines; (iv) agrocin sensitivity; (v) conjugate transfer of Ti plasmids; and (vi) catabolism of arginine and ornithine (Zaenen et al., 1974; Van Larebeke et al., 1974; Watson et al., 1975; Bomhoff et al., 1976; Genetello et al., 1977; Kerr et al., 1977; Guyon et al., 1980; Van Larebeke et al., 1975; Schell, 1975; Engler et al., 1975; Petit et al., 1978a, 1978b; Firmin and Fenwick, 1978; Klapwyk et al., 1978; Ellis et al., 1979). The crown gall tumors contain a DNA segment (called T-DNA) derived from Ti plasmids which is homologous and colinear with a defined fragment of the corresponding Ti plasmid present in the tumor-inducing bacterium (Chilton et al., 1977; Schell et al., 1979; Lemmers et al., 1980; Thomashow et al., 1980; De Beuckeleer et al., 1981). The T-DNA is covalently linked to plant DNA (Zambryski et al., 1980; Yadav et al., 1980; Thomashow et al., 1980) in the nucleus of the plant cell (Chilton et al., 1980; Willmitzer et al., 1980). The T-DNA is transcribed in the transformed plant cell and T-DNA-encoded proteins, such as the octopine synthesizing enzyme, lysopine dehydrogenase (LpDH), have been found in the several octopine crown gall lines (Schroder et al., 1981). This has led to the realization that Ti plasmids are a natural gene vector for plant cells evolved by and for the benefit of the bacteria that harbor Ti plasmids (Schell et al., 1979). All the genetic information for the synthesis of opines in transformed plant cells and for their catabolism by free-living agrobacteria, is carried by Ti plasmids. In this way, free-living agrobacteria can utilize the opines produced by the tumors they have incited as sources of carbon and nitrogen.

In this chapter, we want to concentrate on those aspects of the structure and the properties of Ti plasmids that are important for their use as experimental gene vectors.

INTEGRATION OF A SEGMENT OF THE Ti PLASMID (T-REGION) IN PLANT NUCLEAR DNA

The T-region is defined as that segment of the Ti plasmids that is homologous to sequences present in crown gall cells. The sequences which are transferred from the Ti plasmid to the plant and determine tumorous growth have been called T-DNA. The T-region of octopine and nopaline Ti plasmids have been studied in great detail both physically and functionally. The T-regions, roughly 23 kb in size, are only a portion of the entire plasmids (Lemmers et al., 1980; Thomashow et al., 1980; De Beukeleer et al., 1981; Engler et al., 1981). Southern blotting and cross-hybridization of restriction endonuclease digests of the two types of plasmids as well as electron microscope heteroduplex analyses have revealed that 8 to 9 kb of the T-DNA regions are conserved and common to both octopine and nopaline types of plasmids (Engler et al., 1981; Chilton et al., 1978; Depicker et al., 1978). DNA sequence data confirm that these "common" or "core" segments of the T-regions are about 90% homologous with one another (unpublished results). It was postulated early on (Chilton et al., 1978; Depicker et al., 1978; Schell and Van Montagu, 1977) that this common core might contain genes essential for tumor formation and maintenance. Recent data on the expression of the T-DNA and the study of the effects of insertion and deletion mutations in the common core of the T-region, have verified and extended this hypothesis (see following sections). Attempts to reveal homology between the T-DNA region and plant DNA have failed thus far.

Detailed analysis of some nopaline lines (Lemmers et al., 1980; Zambryski et al., 1980, 1982; Yadav et al., 1980) suggests that the mechanism of T-DNA integration is rather precise since the same continuous segment of the Ti plasmid is always present. Some lines appear to contain a single T-DNA copy, whereas in others the T-DNA occurs in multiple copies which are organized in a tandem array.

Several octopine tumor lines have been studied, and the data suggest that the octopine T-DNA is more variable (Thomashow et al., 1980a and b; De Beuckeleer et al., 1981). A left T-DNA region (TL) containing the "common" or "core" sequence is always present; this region is usually 12 kb in size but one Petunia tumor line is shortened at the right end of TL by about 4 kb (De Beukeleer et al., 1981). In addition, there often is a right T-DNA region (TR) which contains sequences which are adjacent but not contiguous in the octopine Ti plasmid. In some tumor lines TR is amplified whereas TL is not (Thomashow et al., 1980; Merlo et al., 1980). Recent observations demonstrate that TL can also be part of a tandem array (Holsters et al., 1983). It is not known whether the integration is the result of plant or Ti plasmid-specific functions but it is likely that both are involved.

In view of the observed involvement of the "ends" of the T-region in the integration of T-DNA, it was expected that any DNA segment inserted between these "ends" would be cotransferred, provided no function essential for T-DNA transfer and stable maintenance was inactivated by the insertion. The genetic analysis of the T-region by transposon insertion provided Ti plasmid mutants to test this

hypothesis. A Tn7 insertion in the nopaline synthase locus produced a
Ti plasmid able to initiate T-DNA transfer and tumor formation (De
Beuckeleer et al., 1978; Van Montagu and Schell, 1979; Hernalsteens et
al., 1980).

Analysis of the DNA extracted from these tumors showed that the T-
region containing the Tn7 sequence had been transferred as a single 38
x 10^3-bp-segment without any major rearrangements. Several different
DNA sequences have since been introduced into different parts of the T-
region of octopine and nopaline Ti plasmids (Leemans et al., 1981,
1982; Joos et al., 1983). The preliminary observations with these
mutant Ti plasmids fully confirm our initial expectations: as a rule,
DNA sequences inserted between the "ends" of the T-region are
cotransferred with, and become a stable part of the T-DNA of the plant
tumor cells transformed with these mutant Ti plasmids. If the
experimental insert would inactivate a function essential for the
transfer of the T-region, or for the integration of the T-DNA, such a
mutant Ti plasmid would not be able to transform plant cells. No such
inserts have as yet been characterized, indicating that T-DNA transfer
and integration is probably coded for by genes outside of the T-region,
or by a combination of functions with different genetic localization
(Leemans et al., 1982; Joos et al., 1977).

These observations have, therefore, firmly established that the Ti
plasmids can be used as experimental gene vectors and that large DNA
sequences (up to 50 kb) can be transferred stably to the nucleus of
plant cells as a single DNA segment.

MODIFICATION OF Ti PLASMIDS WITH THE PURPOSE TO IDENTIFY THE TUMOR-CONTROLLING PROPERTIES

A double approach was used to elucidate the mechanism of tumor
formation resulting from the integration of T-DNA in the plant cell
nucleus. It was first shown that the T-DNA consists of a number of
well-defined transcriptional units, which were carefully located on
the physical map of the corresponding T-region. Subsequently,
insertions and deletion substitutions were introduced at specific
sites of the T-region to produce Ti plasmids carrying mutant T-regions
such that one or another, or a combination of several transcripts could
not be formed in plant cells carrying such mutant T-DNAs.

RNA transcripts homologous to T-DNA have been shown to be present in
all crown gall tissues studied thus far (Drummond et al., 1977;
Willmitzer et al., 1981; Gelvin et al., 1981). The number, sizes, and
location of the transcribed T-DNA segments were studied in both
octopine and nopaline tumors (Willmitzer et al., 1982, 1983). Tumor-
specific RNAs were detected and mapped by hybridization of [32]P-
labeled-Ti plasmid fragments to RNA which had been separated on
agarose gels, and then transferred to DBM paper. The results show that
octopine tumors contain a total of twelve distinct transcripts (eight
different transcripts are derived from TL-DNA, and four from TR-DNA),
whereas nopaline tumors were found to contain at least 13 different
transcripts.

These transcripts differ in their relative abundance, and in their sizes. They all bind to oligo(dT)-cellulose, indicating that they are polyadenylated. Thus, the T-DNA transferred from a prokaryotic organism provides specific poly(A) addition sites. The direction of transcription was determined, and the location of the approximate 5'- and 3'-ends were mapped on the T-DNA.

All RNAs mapped within the T-DNA sequence. This, and the observation that transcription is inhibited by low concentrations of alpha-amanitin (Willmitzer et al., 1981) suggests that each transcript is determined by a specific promoter site on the T-DNA recognized by plant RNA polymerase II. Since not all transcripts were synthesized from the same strand of DNA, the simplest model for transcription would be that there is one promoter site per group of transcripts. If so, one would expect that the deletion of a 5'-proximal gene of a group would also lead to the disappearance of the transcripts from the 3'-distal genes. However, analysis of cell lines containing the T-DNA of Ti plasmid mutants indicated that genes could not be inactivated by mutations lying far from the coding region (Leemans et al., 1982; Joos et al., 1983). The results available so far are consistent with the assumption that each gene on the T-DNA has its own signals for transcription in the eukaryotic plant cells. Six transcripts were found to be derived from the "common" or "core" segment of the T-region. These transcripts were found to be identical in nopaline and octopine tumors (Willmitzer et al., 1982, 1983).

In order to determine whether these mRNAs are translated into proteins, a hybridization selection procedure was developed that was sufficiently sensitive and specific to detect mRNAs which represent about 0.0001% of the total mRNA activity in the plant cell (the concentration of total T-DNA-specific RNA in the octopine tumor line A6-S1 is between 0.0005 and 0.001%). This procedure was used to enrich for T-DNA-derived mRNAs by hybridization to Ti plasmid fragments covalently bound to microcrystalline cellulose; the hybridized RNAs were eluted and translated in vitro in a cell-free system prepared from wheat germ.

The results obtained with this approach (Schroder and Schroder, 1982) showed that tumor cells contain at least three T-DNA-derived mRNAs which can be translated in vitro into distinct proteins. The protein encoded at the right end of the TL-DNA (molecular weight 39,000 dalton) was of particular interest since previous genetic analysis indicates that this region is responsible for octopine synthesis (Koekman et al., 1979; De Greve et al., 1981; Garfinkel et al., 1981).

The in vitro synthesized protein was shown to be identical in size with the octopine-synthesizing enzyme in octopine tumors. Immunological studies showed that this protein was recognized by antiserum against the tumor-specific synthase (Schroder et al., 1981). These results demonstrate that the structural gene for the octopine-synthesizing enzyme is on the Ti plasmid. So far, this is the only protein product of the T-DNA with known enzymatic properties; the possible functions of two smaller T-DNA-derived proteins are not known. The region coding for the octopine-synthesizing enzyme has recently been sequenced (De Greve et al., 1982). The 5'-end of the octopine synthase mRNA was accurately mapped by sequencing a T-region DNA fragment that hybridizes to this mRNA and thus protects it from degradation by the single strand-specific S1 nuclease. The promoter sequence thus identified is more eukaryotic than prokaryotic in its recognition signals, and no introns interrupt the open-reading frame which starts at the first AUG codon following the 5'-start of the transcript. Similar work leading to essentially the same conclusions has also been performed for the nopaline synthase (Depicker et al., 1982). It is important to note that the promoter sequences for the octopine synthase appear to escape possible control mechanisms since they remain active in all tissues of plants regenerated from tobacco cells transformed with Ti plasmids (De Greve et al., 1982).

The 3'-polyadenylated terminus of the transcript was also analyzed, and a polyadenylation signal 5'-AAUAA-3' was found about 10 nucleotides from the start of the poly(A) sequence. This appears to be a general feature of eukaryotic mRNAs since it has been observed in a number of animal mRNAs (Fitzgerald and Shenk, 1981). To some extent, therefore, these opine-synthesizing genes seem designed to function in eukaryotic cells rather than in prokaryotic cells.

However, this is not necessarily true for all genes of the T-region, since transcripts were also detected in agrobacteria (Gelvin et al., 1981; unpublished data). For this reason, it was interesting to determine whether all the T-DNA-derived mRNAs isolated from plant cells shared properties with typical eukaryotic mRNAs. The fact that translation of each of the three mRNAs analyzed in vitro was inhibited by the cap analogue pm$_7$G suggests that they contain a cap structure at the 5'-end. This would be typical for eukaryotic mRNA since caps have not been described in prokaryotic RNA.

As noted above, the mRNAs of the three in vitro synthesized proteins each represent about 0.0001% of the total mRNA activity in polyribosomal RNA, and this appears to be the detection limit at present for translatable RNA. Some of the other transcripts detected by hybridization experiments are present at even lower concentrations. Assuming that they possess mRNA activity, this is likely to be the reason why the corresponding proteins could not be identified so far by in vitro translation.

A different approach has been developed to search for coding regions on the T-DNA and their protein products. Fragments from the T-region were cloned into E. coli plasmids and analyzed for gene expression in E. coli minicells (Schroder et al., 1981). There are at least four different coding regions within the TL-DNA that can be expressed from promoters which are active in prokaryotic cells and translated into proteins in minicells. The four regions expressed in E. coli correlated with four regions transcribed into RNA in plant cells. The plant transcripts are larger than the proteins in E. coli, and the regions expressed in minicells appear to lie within the regions transcribed in plant cells. One can, therefore, speculate that plant cells and E. coli, at least partly, express the same coding regions.

Specific mutations were introduced in the T-DNA regions of octopine and nopaline Ti plasmids to produce transformed plant cells in which one or more T-DNA-derived transcripts would not be expressed.   By observing the phenotypes of the plant cells harboring such partially inactivated T-DNAs, it was possible to assign functions to most of the different transcripts (Leemans et al., 1982; Joos et al., 1983).   It was found that none of the T-DNA transcripts was essential for the transfer and stable maintenance of T-DNA segments in the plant genome. Essentially, two different functions were found to be determined by T-DNA transcripts.

(i) Transcripts coding for opine synthase: octopine tumors contain either one or three of such genes.   One of them is located on the right end of TL, and codes for octopine synthase, the others are located at the right end of TR and codes for agropine and mannopine synthase (J. Velten, and J. Ellis, personal communication).   Tumors that contain both TL and TR therefore produce both octopine and agropine.   Nopaline tumors also contain at least two transcripts coding for different opines (Willmitzer et al., 1983).   One is located at the right end of the T-DNA and codes for nopaline synthase, whereas the other is located in the left part of the T-DNA, and codes for agrocinopine.

(ii) Transcripts (probably after translation into proteins) that are directly or indirectly responsible for tumorous growth: these transcripts are found to be derived from the "common" or "core" region of the T-DNA.   In total six different well-defined transcripts were found to be derived from this "common" region.   Remarkably, all T-DNA functions affecting the tumor phenotype were located in this "common" region of the T-DNA.   Several of these transcripts act by suppressing plant organ development.   It was observed that shoot and root formation are suppressed independently and by different transcripts.   Two transcripts (1 and 2) were identified that specifically prevent shoot formation.   The effect of these T-DNA gene products is in many ways analogous to that of auxin-like plant growth hormones since the effect of these genes is similar to that observed for calli from normal plant cells with artificially increased auxin level.   Another transcript (transcript 4) was found to prevent specifically root formation, and the effect of this T-DNA gene can, therefore, be compared to the effects observed when normal plant cells are grown in the presence of high concentrations of cytokinins (Leemans et al., 1982; Joos et al., 1983).

That both the shoot and root inhibition resulting from the activity of these genes may be due to the fact that they directly or indirectly determine the formation of auxin- and cytokinin-like growth hormones (Skoog and Miller, 1957; Ooms et al., 1981), is further substantiated by our observation that these genes respectively inhibit shoot or root formation both in T-DNA-containing and T-DNA-negative (normal) cells, provided both types of cells grow as one mixed tissue.

This interpretation of the possible function of transcripts 1 and 2 (auxin-like) and of transcript 4 (cytokinin-like) is consistent with recent measurement of endogenous levels of auxin and cytokinin in teratoma and unorganized crown gall tissue (Amasino and Miller, 1982).

It is important to note that the products of genes 1 and 2 not only suppress   shoot   formation   but   also   stimulate   root   formation.

Reciprocally, the product of gene 4 not only inhibits root formation, but also stimulates shoot formation. This conclusion is based on recent unpublished observations in the authors' laboratories indicating that Ti plasmids from which genes 1 and/or 2 as well as gene 4 were eliminated by deletion/substitution, could transfer their modified T-DNA to plant cells, but did not promote either shoot or root development at the site of infection. Shoot formation therefore requires both the inactivity of genes 1 and/or 2, and the activity of gene 4, whereas root formation requires inactivity of gene 4 and activity of genes 1 and 2. Whereas these observations are consistent with the idea that the products of genes 1, 2 and 4 directly determine the auxin-cytokinin levels in the transformed cells, and that these hormone levels in turn would be responsible for the observed tumor phenotypes, they do not prove this point. It is, for instance, still conceivable that the products of genes 1, 2 and 4 could act directly at the level of gene regulation, and that the alterations in growth hormone levels would be the consequence, rather than the cause, of the observed tumor morphology. It is essential to isolate these gene products in order to determine in detail their mechanism of action. In addition to this hormone-like activity, the T-DNA codes for at least three other transcripts, 5, 6 and 6a. The combination of two of these transcripts, 6 and 6a, with the cytokinin-like activity of transcript 4 was shown to be sufficient to suppress development of transformed cells and to allow their hormone-independent growth (Leemans et al., 1982). Another transcript, 5, was found to inhibit the organization of transformed cells into leaf bud structures. Elimination of this transcript, along with the shoot-inhibiting auxin-like genes (genes 1 and 2) resulted in transformed cells organizing themselves as teratomas (Leemans et al., 1982; Joos et al., 1983).

Whereas the hormone-like effect of genes 1, 2, and 4 results in a suppression of regeneration by both non-T-DNA-containing as well as by T-DNA-containing plant cells, the effect of gene 5 seems to be restricted to the plant cells in which this gene is present and active.

Further evidence in favour of the idea that the "onc" genes of the T-DNA are responsible for tumor formation primarily because they negatively control (suppress) differentiation of shoots and roots is based on an analysis of spontaneous deletion mutants. It is observed that untransformed cells, mixed in primary tumors with T-DNA-containing transforming cells, are able to regenerate normal plants provided that the genes 1 and/or 2 of the transformed cell are inactive. The transformed cells themselves are still suppressed for regeneration, but it seems reasonable that, if spontaneous mutations would inactivate the genes responsible for the suppression of the transformed cells, then such cells would also be able to regenerate and form shoots and roots. In order to recognize such plants derived from cells containing mutated T-DNAs, a large number of shoots from a shooting octopine tumor were screened for the presence of octopine synthase. Most of the shoots were negative, but some of the proliferating shoots were positive. Several of these shoots were grown further on growth-hormone-free media, and found to develop roots, and later to grow into fully normal, flowering plants. Each part of these plants: leaves, stem, and roots were found to contain octopine synthase activity, and polysomal RNA was found to contain T-DNA transcripts homologous to the opine synthesis locus.   No transcripts of the common segment of the T-region were observed.   One of these plants, rGV1, was studied in great detail (De Greve et al., 1982), and its T-DNA was isolated from the plant DNA by molecular cloning in a λ phage vector.   The T-DNA was found to have undergone a large deletion removing all but the right-most part of the T-DNA which codes for the octopine synthase gene.   This explains the transcription data of LpDH-positive plants.   Based on this tissue line, it would appear that for fully normal plants to be formed by T-DNA-containing cells, it is essential that genes 1, 2, and 4, and possibly genes 5, 6a and 6b are inactivated.

We are uncertain about genes 6a and 6b because no important phenotypic change has thus far been correlated with their inactivation (Joos et al., 1983).

An apparently different conclusion could result from recent observations by Barton and Chilton (personal communication).   These authors found that they could regenerate normally organized plants from nopaline-positive roots obtained by cellular cloning of a rooting tumor.   This rooting tumor was induced by a nopaline T-region mutant in which only gene 4 is inactive.   These "normal" nopaline-positive plants were analyzed by DNA/DNA hybridizations, and were found to contain all of the mutated T-region.   In fact, the T-DNA appeared to be amplified to twenty copies per genome in these plants.

Since no transcription studies have yet been done with these plants, it is still conceivable that the other genes of the common T-region are inactive in these plants due to a mutation other than a deletion which would not be detected by Southern gel blotting DNA/DNA hybridizations.   Alternatively, the prior inactivation of gene 4 could create the proper conditions for regeneration even in the presence of genes 5, 1, and 2.   These observations together with ours (Leemans et al., 1982) showing that hormone-independent, teratoma-like tumorous growths can be obtained by the presence solely of genes 4, 6a, and 6b (preliminary results indicate that T-DNAs with gene 4 as the only active gene strongly induce shoot and teratoma formation possibly indicating that genes 6a and 6b do not have a qualitative but possibly a quantitative effect, e.g. on the expression of gene 4), indicate that the product of gene 4 is probably the most important for the formation and the maintenance of the tumorous state in Ti plasmid-transformed plant cells.

Support for this hypothesis comes from the observation (J. Tempe, personal communication) that hairy roots of carrots and cauliflower, induced by infection with A. rhizogenes, can be regenerated easily into plants since these hairy roots have a T-region which is homologous with gene 1 coding for shoot suppression in octopine and nopaline tumors, but do not contain a gene equivalent to gene 4 (Willmitzer et al., 1982).

The question can be asked whether the general functional organization of the T-DNA-linked genes, as studied for octopine and nopaline crown gall tumors, also applies to other types of tumors or transformations

induced by agrobacteria. Agropine (previously null-type) tumors
(Guyon et al., 1980) have a T-region which is homologous, based on
DNA/DNA hybridization studies (Drummond and Chilton, 1978), with the
genes coding for transcripts 5, 2, 1, and 4 of the "common" region of
octopine and nopaline T-regions. No homology was found to transcripts
6a and 6b.

In the previous literature observations are described suggesting that
T-DNA could not pass through meiosis. Seeds obtained by self-
fertilization of LpDH-positive plants, however, produced new plants
with active T-DNA-linked genes, demonstrating that genes introduced in
plant nuclei, via the Ti plasmid, can be sexually inherited (Otten et
al., 1981). A series of sexual crosses was therefore designed to study
the transmission pattern of the T-DNA-specific genes. The results of
these crosses demonstrate very convincingly that the T-DNA-linked
genes (LpDH) are transmitted as a single Mendelian factor both through
the pollen and through the eggs of the originally transformed plant.
These crosses also showed that the original transformed plant was a
hemizygote containing T-DNA only on one of a pair of homologous
chromosones. By these crosses tobacco plants homozygotic for the
altered T-DNA were obtained (Otten et al., 1981). When regenerants
from different transformations are crossed the two T-DNA loci
segregate independently (De Greve et al., 1982). Subsequent
experiments have demonstrated that mutant Ti plasmids reproducibly
give rise to normal plants in tobacco, petunia, and potato. In all
these cases, the plants were shown to contain and express the octopine
and/or agropine synthase genes of the mutant T-DNA.

DEVELOPMENT OF A PRACTICAL PLANT HOST VECTOR

The evidence described above led to the notion that the genes carried
by the T-region of Ti plasmids are not essential for either transfer of
the plasmid DNA or for its stable integration into the chromosomal DNA
of plant cells. On the other hand, the so-called "Vir"-region and the
"border" or "recognition" sequences (the latter sequences are located
to the left an the right of the T-region) were shown to be important
for either transfer or integration.

On the basis of these notions, it became possible to design a vector
system for practical use. Such a vector system consists of two
components: (i) an "intermediate vector" which can be one of the
commonly used cloning vehicles, such as pBR322, into which "foreign"
genes can be cloned and which can easily be mobilized to Agrobacterium
by conduction; and (ii) an "acceptor" Ti plasmid which carries an
active "Vir"-region and the "borders" or "recognition" sequences
involved in integration specificity. Located in-between these
"border" or "recognition" sequences one would want a DNA sequence
homologous to the cloning vector (e.g. pBR322) and ideally a dominant
selectable marker gene for plant cells. By a single homologous
recombination event the vector carrying the foreign gene can thus be
inserted next to the selectable marker gene and in-between the
"border" or "recognition" sequences of the "acceptor" Ti plasmid.

Such a versatile acceptor Ti plasmid (pGV3850) was recently
constructed (Zambryski et al., 1983 submitted) and it was shown that

agrobacteria carrying such a non-oncogenic Ti plasmid vector would transfer and integrate the DNA of the vector into plant cells with remarkable efficiency. Calli derived from wounded plantlets and grown on media promoting differentiation produced normal shoots, and it was found that up to 70% of these shoots could consist of cells containing the DNA from the cloning vehicle and expressing nopaline synthase which was used as a marker gene. These "transformed" shoots readily grew to form normal plants.

A further development in this approach consisted in designing "intermediate expression" vectors that would promote the transcription and translation of foreign genes in plants. Indeed, this became essential when it was observed that a number of foreign genes of bacterial, plant, and animal origin, introduced into tobacco or sunflower via Ti plasmids, were not expressed. This was in sharp contrast to the wild-type opine synthase genes, such as octopine and nopaline synthase, that were shown to be expressed after their transfer in a wide variety of plant cells. These genes were therefore very useful to study the expression of foreign genes in plants. To do this, an intermediate expression vector was constructed (Herrera-Estrella et al., 1983) with an unique BamHI cloning site at the end of the 5'-untranslated leader sequence of the nopaline synthase gene. A large part of the coding sequence of the nopaline synthase gene was removed but the 3'-untranslated region including the polyadenylation signal sequence, was kept. Coding sequences derived from a number of bacterial genes were inserted into this expression vector, such as the neomycin phosphotransferase from Tn5, the methotrexate-insensitive dihydrofolate reductase of the plasmid R67, and the chloramphenicol acetyltransferase from pBR325.

These chimeric genes were transferred along with the intermediate expression vector to tobacco cells, and all three of these chimeric genes were shown to be expressed in the plant cells. Evidence was also obtained that these chimeric genes can be used as potent dominant selectable marker genes (Herrera-Estrella, 1983). Antibiotics such as kanamycin and G418, and drugs such as methotrexate, are very toxic to plant cells. Therefore, the functional expression of the neomycin phosphotransferase and of the methotrexate-insensitive dihydrofolate reductase in plant cells should make the transformed plant cells resistant to kanamycin and G418 on the one hand, and to methotrexate on the other hand. This was indeed shown to be the case.

GENERAL CONCLUSIONS

Considerable efforts will still have to be made before the structure, function, and regulation of plant genes are understood in any detail. The recent advances achieved with host vectors based on the Ti plasmid indicate, however, that very fast progress can be expected.

It is even conceivable that the availability of dominant selectable marker genes will open the way for the development of host gene vectors for monocotyledonous plants which have not yet been shown to be transformable by agrobacteria. The initial phase of this research, with the aim to find out whether and how genes could be transferred to, and expressed in plants, has now been successfully concluded.

REFERENCES

Amasino, R.M. and Miller, C.O. (1982). Plant Physiol. 69, 389-392.

Bomhoff, G., Klapwijk, P.M., Kester, H.C.M., Schilperoot, R.A., and Rorsch, A. (1976). Mol. Gen. Genet. 145, 177-181.

Chilton, M.D., Drummond, H.J., Merlo, D.J., Sciaky, D., Montoya, A.L., Gordon, M.P. and Nester, E.W. (1977). Cell 11, 263-271.

Chilton, M.D., Drummond, H.J., Merlo, D.J. and Sciaky, D. (1978). Nature 275, 147-149.

Chilton, M.D., Saiki, R.K., Yadav, N., Gordon, M.P. and Quetier, F. (1980). Proc. Natl. Acad. Sci. USA 77, 4060-4064.

Currier, T.C. and Nester, E.W. (1976). J. Bacteriol. 126, 157-165.

De Beuckeleer, M., De Block, M., De Greve, H., Depicker, A., De Vos, R., De Vos, G., De Wilde, M., Dhaese, P., Dobbelaere, M.R., Engler, G., Genetello, C., Hernalsteens, J.P., Holsters, M., Jacobs, A., Schell, J., Seurinck, J., Silva, B., Van Haute, E., Van Montagu, M., Van Vliet, F., Villarroel, R. and Zaenen, I. (1978). Proceedings IVth International Conference on Plant Pathogenic Bacteria (M. Ride, ed.), pp. 115-126. I.N.R.A., Angers.

De Beuckeleer, M., Lemmers, M., De Vos, G., Willmitzer, L., Van Montagu, M. and Schell, J. (1981). Mol. Gen. Genet. 183, 283-288.

De Greve, H., Decraemer, H., Seurinck, J., Van Montagu, M. and Schell, J. (1981). Plasmid 6, 235-248.

De Greve, H., Dhaese, P., Seurinck, J., Lemmers, M., Van Montagu, M. and Schell, J. (1982). J. Mol. Appl. Genet. 1, 499-512.

De Greve, H., Leemans, J., Hernalsteens, J.P., Thia-Toong, L., De Beuckeleer, M., Willmitzer, L., Otten, L., Van Montagu, M. and Schell, J. (1982). Nature 300, 752-755.

Depicker, A., Stachel, S., Dhaese, P., Zambryski, P. and Goodman, H.M. (1982). J. Mol. Appl. Genet. 1, 561-574.

Depicker, A., Van Montagu, M. and Schell, J. (1978). Nature 275, 150-153.

Drummond, M.H. and Chilton, M.-D. (1978). J. Bacteriol. 136, 1178-1183.

Drummond, M.H., Gordon, M.P., Nester, E.W. and Chilton, M.-D. (1977). Nature (London) 269, 535-536.

Ellis, J.G., Kerr, A., Tempe, J. and Petit, A. (1979). Mol. Gen. Genet. 173, 263-269.

Engler, G., Depicker, A., Maenhaut, R., Villarroel-Mandiola, R., Van Montagu, M. and Schell, J. (1981). J. Mol. Biol. 152, 183-208.

Engler, G., Holsters, M., Van Montagu, M., Schell, J., Hernalsteens, J.P. and Schilperoort, R.A. (1975). Mol. Gen. Genet. 138, 345-349.

Firmin, J.L. and Fenwick, G.R. (1978). Nature 276, 842-844.

Fitzgerald, M. and Shenk, T. (1981). Cell 24, 251-260.

Garfinkel, D.J., Simpson, R.B., Ream, L.W., White, F.F., Gordon, M.P. and Nester, E.W. (1981). Cell 27, 143-153.

Gelvin, S.B., Gordon, M.P., Nester, E.W. and Aronson, A.I. (1981). Plasmid 6, 17-29.

Genetello, C., Van Larebeke, N., Holters, M., Depicker, A., Van Montagu, M. and Schell, J. (1977). Nature 265, 561-563.

Gurley, W.B., Kemp, J.D., Albert, M.J., Sutton, D.W. and Callis, J. (1979). Proc. Natl. Acad. Sci. USA 76, 2828-2832.

Guyon, P., Chilton, M.-D., Petit, A. and Tempe, J. (1980). Proc. Natl. Acad. Sci. USA 77, 2693-2697.

Hernalsteens, J.P., Van Vliet, F., De Beuckeleer, M., Depicker, A., Engler, G., Lemmers, M., Holsters, M., Van Montagu, M. and Schell, J. (1980). Nature 287, 654-656.

Herrera-Estrella, L., De Block, M., Messens, E., Hernalsteens, J.-P., Van Montagu, M. and Schell, J. (1983). EMBO J. 2, 987-995.

Herrera-Estrella, L., Depicker, A., Van Montagu, M. and Schell, J. (1983) Nature 303, 209-213.

Holsters, M., De Waele, D., Depicker, A., Messens, E., Van Montagu, M. and Schell, J. (1978). Mol. Gen. Genet. 163, 181-187.

Holsters, M., Villarroel, R., Gielen, J., Seurinck, J., De Greve, H., Van Montagu, M. and Schell, J. (1983). Mol. Gen. Genet. 190, 35-41.

Hooykass, P.J.J., Klapwijk, P.M., Nuti, M.P., Schilperoot, R.A. and Rorsch, A. (1977). J. Gen. Microbiol. 98, 477-484.

Joos, H., Inze, D., Caplan, A., Sormann, M., Van Montagu, M. and Schell, J. (1983). Cell 32, 1057-1067.

Kerr, A. (1969). Nature 223, 1175-1176.

Kerr, A. (1971). Physiol. Plant Pathol. 1, 241-246.

Kerr, A., Manigault, P. and Tempe, J. (1977). Nature 265, 560-651.

Klapwijk, P.M., Scheulderman, T. and Schilperoort, R.A. (1978). J. Bacteriol. 136, 775-785.

Koekman, B.P., Ooms, G., Klapwijk, P.M. and Schilperoort, R.A. (1979). Plasmid 2, 347-357.

Leemans, J., Deblaere, R., Willmitzer, L., De Greve, H., Hernalsteens, J.P., Van Montagu, M. and Schell, J. (1982). EMBO J. 1, 147-152.

Leemans, J., Shaw, C., Deblaere, R., De Greve, H., Hernalsteens, J.P., Maes, M., Van Montagu, M. and Schell, J. (1981). J. Mol. Appl. Genet. 1, 149-164.

Lemmers, M., De Beuckeleer, M., Holsters, M., Zambryski, P., Depicker, A., Hernalsteens, J.P., Van Montagu, M. and Schell, J. (1980). J. Mol. Biol. 144, 353-376.

Merlo, D.J. and Nester, E.W. (1977). J. Bacteriol. 129, 76-80.

Merlo, D.J., Nutter, R.C., Montoya, A.L., Garfinkel, D.J., Drummond, M.H., Chilton, M.-D., Gordon, M.P. and Nester, E.W. (1980). Mol. Gen. Genet. 177, 637-643.

Ooms, G., Hooykaas, P.J., Moleman, G. and Schilperoort, R.A. (1981). Gene 14, 33-50.

Otten, L., De Greve, H., Hernalsteens, J.P., Van Montagu, M., Schieder, O., Straub, J. and Schell, J. (1981). Mol. Gen. Genet. 183, 209-213.

Petit, A., Dessaux, Y. and Tempe, J. (1978a). in Proceedings IVth International Conference on Plant Pathogenic Bacteria (M. Ride, ed.), pp. 143-152, I.N.R.A., Angers.

Petit, A., Tempe, J., Kerr, A., Holsters, M., Van Montagu, M. and Schell, J. (1978b). Nature 271, 570-572.

Schell, J. (1975). In Genetic Manipulations with Plant Materials, (L. Ledoux, ed.), pp. 163-181, Plenum Press, New York.

Schell, J. and Van Montagu, M. (1977). Brookhaven Symp. Biol. 29, 36-49.

Schell, J., Van Montagu, M., De Beuckeleer, M., De Block, M., Depicker, A., De Wilde, M., Engler, G., Genetello, C., Hernalsteens, J.P., Holsters, M., Seurinck, J., Silva, B., Van Vliet, F. and Villarroel, R. (1979). Proc. Roy. Soc. Lond. B 204, 251-266.

Schroder, G. and Schroder, J. (1982). Mol. Gen. Genet. 185, 51-55.

Schroder, J., Hillebrandt, A., Klipp, W. and Puhler, A. (1981). Nucl. Acids Res. 9, 5187-5202.

Schroder, J., Schroder, G., Huisman, H., Schilperoort, R.A. and Schell, J. (1981). FEBS Lett. 129, 166-168.

Skoog, F. and Miller, C.O. (1957). Symp. Soc. Exp. Biol. 11, 118-131.

Thomashow, M.F., Nutter, R., Montoya, A.L., Gordon, M.P. and Nester, E.W. (1980a). Cell 19, 729-739.

Thomashow, M.F., Nutter, R., Postle, K., Chilton, M.-D., Blattner, F.R., Powell, A., Gordon, M.P. and Nester, E.W. (1980b). Proc. Natl. Acad. Sci. USA 77, 6448-6452.

Van Larebeke, N., Engler, G., Holsters, M., Van den Elsacker, S., Zaenen, I., Schilperoort, R.A. and Schell, J. (1974). Nature 252, 169-170.

Van Larebeke, N., Genetello, C., Hernalsteens, J.P., Depicker, A., Zaenen, I., Messens, E., Van Montagu, M. and Schell, J. (1977). Mol. Gen. Genet. 152, 119-124.

Van Larebeke, N., Genetello, C., Schell, J., Schilperoort, R.A., Hermans, A.K., Hernalsteens, J.P. and Van Montagu, M. (1975). Nature 255, 742-743.

Van Montagu, M. and Schell, J. (1979). In Plasmids of Medical, Environmental and Commercial Importance (K. Timmis and A. Puhler, eds), pp. 71-96, Elsevier, Amsterdam.

Watson, B., Currier, T.C., Gordon, M.P, Chilton, M.-D and Nester, E.W. (1975). J. Bacteriol. 123, 255-264.

Willmitzer, L., De Beuckeleer, M., Lemmers, M., Van Montagu, M. and Schell, J. (1980). Nature 287, 359-361.

Willmitzer, L., Dhaese, P., Schreier, P.H., Schmalenbach, W., Van Montagu, M. and Schell, J. (1983). Cell 32, 1045-1056.

Willmitzer, L., Otten, L., Simons, G., Schmalenbach, W., Schroder, J., Schroder, G., Van Montagu, M., De Vos, G. and Schell, J. (1981). Mol. Gen. Genet. 182, 255-262.

Willmitzer, L., Schmalenbach, W. and Schell, J. (1981). Nucl. Acids Res. 9, 4801-4812.

Willmitzer, L., Sanchez-Serrano, J., Buschfeld, E. and Schell, J. (1982). Mol. Gen. Genet. 186, 16-22.

Willmitzer, L., Simons, G. and Schell, J. (1982). EMBO J. 1, 139-146.

Yadav, N.S., Postle, K., Saiki, R.K., Thomashow, M.F. and Chilton, M.-D. (1980). Nature 287, 458-461.

Zaenen, I., Van Larebeke, N., Teuchy, H., Van Montagu, M. and Schell, J. (1974). J. Mol. Biol. 86, 109-127.

Zambryski, P., Depicker, A., Kruger, K. and Goodman, H. (1982). J. Mol. Appl. Genet. 1, 361-370.

Zambryski, P., Holsters, M., Kruger, K., Depicker, A., Schell, J., Van Montagu, M. and Goodman, H.M. (1980). Science 209, 1385-1391.

Zambryski, P., Joos, H., Genetello, C., Van Montagu, M. and Schell, J. (1983). Submitted.

VARIABILITY IN TISSUE CULTURE DERIVED PLANTS

Horst Lörz

Max-Planck-Institut für Züchtungsforschung, 5000 Köln 30
Fed. Rep. of Germany

ABSTRACT

Phenotypic and genetic variation in plants regenerated from *in vitro*
cell cultures is found in many different species and for many different characters. This tissue culture variability or somaclonal variation is dependent upon species and genotypes used for *in vitro* culture, the source of explant used for culture initiation and the culture
procedure itself. Both, preexisting variation in the somatic plant
tissue and to a larger extent cell culture-induced variation contribute to generate somaclonal variation. There is substantial evidence
that a significant proportion of somaclonal variation is genetic.
Changes in ploidy level or chromosome number and chromosomal rearrangements are most frequently described and discussed to explain the
origin of variability. At the molecular level, transposition events
have been found by DNA restriction enzyme analysis and also transposable elements have been included in the list of possible mechanisms
involved with somaclonal variation. Genetic analysis of the progeny
from tissue culture-derived plants resulted in the isolation of numerous mutants, predominantly recessive ones. Although tissue culture
induced variability is not yet fully understood and therefore only
partially manageable, application of somaclonal variation is seen already in an agricultural context. In particular,useful variability in
tissue culture derived crop plants is found for disease resistance
and for characters where *in vitro* selection procedures are available
and advantageous compared to selection in conventional plant breeding
programs.

INTRODUCTION

In vitro techniques for plants have progressed in the last two decades to a state where a considerable contribution for plant improvement can be expected (Wenzel et al., Peacock et al., this symposium).
The new techniques certainly will not replace the skillful work of
conventional plant breeders, but have opened up possibilities not
readily attainable by conventional plant breeding procedures. Progress in plant breeding is achieved in general by creating new variability and application of efficient selection for the desired characters. The creation of variablility other than by conventional sexual means is now possible with new techniques of plant biotechnology,
such as somatic hybridization, gene transfer into plant cells (Schell
et al., Riedel, this symposium), *in vitro* pollination and embryo culture, cell and tissue culture and *in vitro* selection (Fujiwara 1982).

In this report special emphasis will be given to variability which is

found among plants regenerated from cell culture, a phenomenon also
called somaclonal variation (Larkin and Scowcroft 1981). A cell culture cycle of higher plants refers to a procedure which, in principle, involves the establishment of *in vitro* cultures from different
sources (multicellular explants or isolated single cells, protoplasts),
the maintenance and proliferation of the cultures as undifferentiated
masses of cells (callus, cell suspension) on defined media, and finally the regeneration of plants after induction of shoot and root
morphogenesis or via somatic embryogenesis.

It is not the intention of this review to enumerate all species where
somaclonal variation has been described, but rather to discuss some
generalities of this phenomenon.

THE PHENOMENON OF VARIABILITY

Originally cell culture was seen as a new procedure of clonal propagation of plants and phenotypically identical regenerants were expected. However, exceptions were found especially in asexually propagated species, such as sugar cane (Heinz et al. 1977), potato (Shepard et al. 1980) and *Pelargonium* (Skirvin 1978). These authors judged the phenotypic variability they found not simply as an experimental artefact but rather introduced the incidence of tissue culture
variability as an introduction of a new option for plant improvement.

Sugar cane tissue culture derived somaclones were identified which
showed resistance to Fiji virus disease and to downy mildew, although
the tissue cultures had been initiated from susceptible cultivars
(Heinz et al. 1977). Analogous results were obtained by Larkin and
Scowcroft (1983) who initiated cell cultures from stem segments of a
line which is highly susceptible to eyespot disease (*Helminthosporium
sacchari*). Among the regenerated plants these authors were able to
isolate somaclones which exhibited different degrees of tolerance.
The incidence of tolerant lines was not limited to a cell culture
procedure where a selection with the fungal toxin was applied, but
somaclones with increased tolerance were found also without any specific selection during cell culture and plant regeneration. In subsequent vegetative generations the resistance to the *Helminthosporium*
toxin was retained in 70 % of the somaclones, 10 % were again sensitive, and the rest segregated. When primary somaclones were used for
a second tissue culture cycle the regenerants showed again variation.
Relative to the primary somaclones, about 20 % showed more resistance,
40 % less resistance, and 40 % retained their level. These results
indicate that at least some of the variation is retained after vegetative propagation and under favourable conditions several new characters derived from tissue culture variability may be combined in
subsequent tissue culture cycles.

Besides sugar cane, potato is one of the crop species where somaclonal variation has been studied intensively (Shepard et al. 1980; Bidney and Shepard 1981). Shepard and coworkers developed a rather complex but efficient protocol for plant regeneration from potato protoplasts. The regenerated protoplast- and single cell-derived plants
(protoclones) were screened for resistance to the pathogens *Alter-*

naria solani and *Phytophthora infestans,* and lines with increased tolerance were found with a frequency of 1 % and 2.5 % respectively.
During generative propagation these characters were retained. In a
detailed analysis under field conditions, thirty five different characters were evaluated in protoplast-derived lines, and significant
variation compared to the original line was found in twnty two characters (Secor and Shepard 1981, Shepard 1981). Most of the 65 somaclones analysed in this study showed variation in four characters,
but lines with up to seventeen significantly varying characters also
were found. Thomas et al. (1982) observed a high degree of variability with another commercial potato variety. None of the 23 protoplast-
derived plants analysed resembled each other in all of the ten morphological characteristics scored in this study, and only one regene-
rant resembled the parental type.

Somaclonal variation is not limited to vegetatively propagated species and has been found in many other species too including the tissue culture model plant species *Nicotiana* (reviews by Larkin and
Scowcroft 1981, Scowcroft and Larkin 1983a and 1983b). Already in
1969, Sacristan and Melchers observed variation among tobacco somaclones. A detailed study was undertaken by Barbier and Dulieu (1980)
who initiated cell cultures from *Nicotiana tabacum* heterozygous for
two loci affecting chlorophyll synthesis and leaf colour. Genetic
analysis of the progeny from somaclonal variants indicated at least
3.5 to 3.6 % mutation frequency for each of the loci. In addition to
leaf colour variants they found unstable chlorophyll variants and variants for leaf shape and size. Convincing evidence for mutational
events in somaclones of tobacco was provided also by Burk and Matzinger (1976). They used dihaploid material derived from anther culture
of a line which was already highly inbred. Selfed progeny of tissue
culture-derived lines were analysed with respect to characters such
as yield, grade index, days of flowering, total alkaloids and many
more. Significant variability was found for all ten characters analysed. It is unlikely that all variation resulted from the release of
residual heterozygosity, but rather from mutational events during
cell culture. In an extended study (Javier et al. 1983), variability
was analysed among field-grown progenies of nonconventionally derived
diploid plants. The sources for these plants were anther culture derived haploids, second cycle haploids from anthers of first cycle doubled haploids, and adventitiously derived plants obtained from leaf
cultures of originally haploid material and from normal diploids.
Regeneration of plants from haploid and diploid leaves introduced
new variability. The new variability found after each cycle of diploid leaf regeneration was similar in magnitude to that found for
anther culture regeneration.

Mutational events could be associated with *in vitro* culture also in
Zea mays (Edallo et al. 1981). In this case immature embryos were
used for initiation of morphogenic maize cultures and fertile plants
were regenerated. Scoring for the presence of simply inherited mutations was done in the second progeny generation (R_2) from tissue culture regenerated plants (R_0). On the mean each regenerated plant was
bearing 0.8 to 1.2 mutations (two different parent lines were used).
The type of mutations were phenotypically similar to spontaneous endosperm and seedling mutants described for maize.

Although only a few examples have been mentioned it is obvious that
somaclonal variation is a common phenomenon. Variability is found for
many different characters, and some of the somaclonal variants are
confirmed to be solid mutants at least in such species where progeny
analysis is possible.

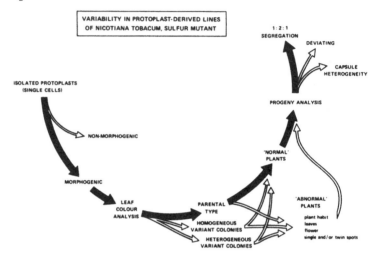

Figure 1:   Occurrence of variability in plants regenerated
from protoplasts of heterozygote sulfur tobacco. The phenotypic and
genetic variability included loss of morphogenetic capacity, variation in leaf colour (sulfur locus), abnormalities in leaf, flower,
and plant morphology, sterility, seed viability, and deviating segregation patterns in the progeny of selfed plants.

VARIATION - PREEXISTING OR CELL CULTURE-INDUCED

The question whether the observed variability reflects genetic variation which existed already in the somatic cells prior to culture initiation or is induced mainly during cell culture is not only of academic interest but has great implications to the possibilities to influence the degree of variability experimentally. This problem was
specifically approached by Lörz and Scowcroft (1983) by regenerating
many plants from protoplasts of Su/su heterozygotes of *Nicotiana tabacum* and monitoring variation at a specific locus. The colony types,
homogeneous or heterogeneous with respect to leaf colour in regenerated shoots, allowed classification into basically three groups: I)
parental type colony, homogeneous colour, heterozygote genotype. II)
homogeneous variant colony, all shoots of the same type, but different
from the parent, III) heterogeneous variant colony, different types of
shoots from one callus originally derived from a single cell. The variant colonies of type II are seen as a result of somatically preexisting variation or, more generally, variation at the single cell level. In contrast, type III colonies are seen as products which arose
from somaclonal variation occurring during cell culture. The leaf colour analysis revealed 92.2 - 98.4 % parental type colonies,

1.4 - 6.0 % cell culture induced variant colonies and only 0.1 -
1.8 % "preexisting" variants. Under all different culture conditions
we found evidence for more "induced" than "preexisting" variability.
Further evidence that the cell culture procedure is most important for
somaclonal variation is found in a positive correlation of extended
cell culture period and increased frequency of variation (Barbier and
Dulieu 1981, Lörz and Scowcroft 1983). In addition, variation was
found in potato between plants regenerated from a single callus which
originally developed from a single protoplast (Thomas et al. 1982).

FACTORS INFLUENCING VARIABILITY

As indicated already, the period of culture is one of the factors affecting the degree of variability. A positive correlation of *in vitro*
culture duration with the frequency of variation was found for specific marker genes in tobacco (Barbier and Dulieu 1980, Lörz and Scowcroft 1983) and also for less specific characters such as tolerance
for the toxin in sugarcane eyespot disease (Scowcroft and Larkin 1983a).
Cytogenetic analysis of meiosis in regenerated oat plants by McCoy et
al. (1982) revealed an increase of cytologically abnormal plants from
12 to 48 % in one line and from 49 to 88 % in another line when the
plants were regenerated after 4 and 20 months in cell culture, respectively. From the results obtained with different species and for different characters it is obvious that variability depends very much on
the period of time between cell culture initiation and plant regeneration. When clonal identity of regenerants is desired, e.g. for propagation of ornamental species, the cell culture period should be kept
as short as possible.

So far there is no convincing evidence about other cell culture procedure dependent factors which influence somaclonal variation. Some differences in the frequency of variants were found by Shepard (1981) after regenerating potato plants from protoplasts which were cultured
with different auxins in the medium, namely NAA or 2,4-D. A "mutagenic"
effect of phytohormones used in the cell culture media was discussed
also by D'Amato (1977) as a possible source for chromosomal variability. In our own experiments with the sulfur mutant of tobacco we compared culture media which differed in the source of nitrogen. Reduced nitrogen of the standard medium was replaced by an amino acid mixture.
Differences with respect to the number of somaclonal variants, however, were not statistically significant and may be counterbalanced
too as cell proliferation was different with the two media. The question about the influence of cell culture procedures on variability
should be accessible for those species where plant regeneration from
single cells is highly efficient and reproducible. It will be of special interest to see if the pathway of regeneration, shoot and root
morphogenesis vs. somatic embryogenesis, is of any importance. Conditions for the induction of somatic embryogenesis have been developed
for numerous species including some cereals (Vasil et al. 1982), and
so far phenotypically or cytologically obvious variants have not been
found (Vasil and Vasil 1981).

Another, to some extent, also accessible factor influencing somaclonal
variation is the type of explant used for culture initiation. In the

work from Barbier and Dulieu (1980), less variation was found in
plants regenerated directly from cotyledon explants than after plant
regeneration from protoplasts isolated from the cotyledons.

A significant impact of the genetic background on the level of somaclonal variation is obvious from a number of publications. Cytogenetic
analysis of plants regenerated from tissue cultures of Avena sativa
(McCoy et al. 1982) revealed significant differences in the level of
variation between the two lines used in these experiments. On an average the line Tippecanoe exhibited twice as many cytological alterations than the line Lodi. The types of abnormalities found in both lines
were chromosome breakages, loss of chromosome segments or total chromosomes, trisomy, monosomy, and interchanges. Genotype dependent differences were found also in protoplast-derived potato plants of two
tetraploid cultivars (Karp et al. 1982). Aneuploidy of regenerated
plants was common for both cultivars but the nature of chromosome variation differed (percentage of plants with normal chromosome set,
range of aneuploidy). The obvious lack of variation amongst plants regenerated from protoplasts of another potato line (Wenzel et al. 1979)
probably is due to the different genotypic background rather than the
dihaploidy of the starting material in these experiments. Genetic variability was found by Prat (1983) in Nicotiana sylvestris after protoplast culture in both an original line and a dihaploid androgenetic
line derived from it. Both lines gave rise to more than 58 % tetraploid plants, but were not significantly different. Also calli regenerating both diploid and tetraploid plants were found from the original and the androgenetic line.

The origin of a somewhat controversial discussion about somaclonal
variation in the last few years can be traced back and explained, at
least partially, by our present knowledge that there is a significant
genotypic influence on the level of tissue culture variability. In
contrast to the results with oat where intensive chromosomal variation
was described, the same authors found a high degree of chromosome stability in maize (McCoy et al. 1982, McCoy and Phillips 1982). The culture procedure for both species was basically identical. No obvious
somaclonal variation was observed also in other species, e.g. Datura
(Schieder 1983). This lack of variants derived from cell culture in
some species also may be indicative of a built-in natural selection
system allowing only relatively "normal" cells to regenerate plants.
Thus Nicotiana tabacum or Solanum tuberosum, amphidiploid and tetraploid species respectively, are characterized to have a large genetic
"buffer capacity". Extensive variation in the genome is possible without disturbing the morphogenetic capacity of the cultured cells. Species without this capacity are probably also affected by somaclonal
variation; however, these abnormalities are not easily detected for
there is a strong bias for normal cells to proliferate and to form
plants.

ORIGIN OF VARIABILITY

As previously mentioned, ploidy changes obviously are a common phenomenon in tissue culture derived plants and observed frequently in potato and tobacco (Karp et al. 1982, Prat 1983). Other karyotype changes include aneuploidy, deletions, translocations and other minor

rearrangements as shown in detailed meiotic analysis by Ogihara (1981) for *Haworthia*, McCoy et al. (1982) for oat and Orton (1980) for barley. Minute chromosomal rearrangements, not easily detectable as cytological changes, can be expected too. Whether or not all of these chromosomal rearrangements cause any phenotypic variation or mutations in the regenerated plants is a question open for speculative discussions. Our findings of increased frequency of twins and single spots in protoplast-derived tobacco plants are interpreted as enhanced somatic crossing over and sister chromatid exchange (Lörz and Scowcroft 1983). The frequency of somatic exchange is similarly increased also after chemical mutagenesis or X-ray treatment (Carlson 1974).

Little information is available about variation at the molecular level of the genome in cultured plant cells. Mitochondrial DNA variation in T-cytoplasmic male sterile (T-cms) maize plants regenerated from tissue culture and selected for resistance to *Helminthosporium maydis* race T has been described by Gengenbach et al. (1981). Restriction enzyme analysis of mitochondrial DNA revealed differences between different tissue culture-derived plants and also differences when compared to normal T or N cytoplasm maize plants. Similar results were described by Kemble et al. (1982) after mitochondrial DNA analysis of maize plants which were not exposed to the fungal toxin during culture and regeneration, but resistant lines were found as a consequence of tissue culture variability (Brettell et al. 1980). Recently Chourey and Kemble (1982) described a specific type of transposition event in maize mitochondrial DNA which was clearly correlated to a change in callus morphology (compact→ friable). The observed DNA variation was limited to two plasmid-like DNA components called S1 and S2. A molecular analysis of the genomic DNA of numerous tissue culture-derived potato plants gave preliminary evidence that rearrangements are existent and undoubtedly detectable by routine procedures of molecular biology (Landsmann, pers. commun.). Heritable quantitative and qualitative changes were observed in the nuclear DNA of doubled-haploid plants of *Nicotiana sylvestris* (De Paepe et al. 1983). Doubled-haploid plants derived from consecutive androgenic cycles contained on the average increasing amounts of total DNA and increasing proportions of highly repeated sequences. Recent progress in the characterization of transposable elements in higher plants (Marx 1983, Starlinger et al. this volume) has given the experimental prerequisite to search also for a possible involvement of these elements in somaclonal variation. Other molecular events discussed in the context of tissue culture variability are gene duplications or depletion and, especially related to variation in disease reaction, the role of cryptic viruses (Larkin and Scowcroft 1981, Chaleff 1981). A better understanding of the origin of variability and the causalities leading to somaclonal variant plants would increase our chance to minimize variation when aiming for clonal identity or enhance or even direct variation for purposes of plant improvement.

CONCLUSION

Tissue culture variability is a widespread phenomenon and there is growing awareness that somaclonal variation may provide useful variability for plant improvement. However, the accessibility (manageability) of somaclonal variation, either to enhance or to reduce variability, is limited, for the events creating variation are poorly understood.

To some extent somaclonal variation may be seen as tissue culture mutagenesis and, therefore has to be compared with all aspects of conventional mutagenesis, including the arguable contribution of induced mutants for better crops. Limitations from this area of application can not be excluded when taking into consideration recent findings by Prat (1983). He found in selfed progenies of protoplast-derived plants without recognizable mutant phenotype differences for qualitative characters. Protoplast culture in general induced a depressive effect on *Nicotiana sylvestris* plants.

Figure 2: Diagrammatic summary of factors affecting somaclonal variation in tissue culture-derived plants.

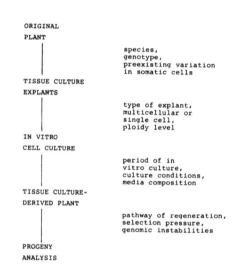

ORIGINAL
PLANT

species,
genotype,
preexisting variation
in somatic cells

TISSUE CULTURE
EXPLANTS

type of explant,
multicellular or
single cell,
ploidy level

IN VITRO
CELL CULTURE

period of in
vitro culture,
culture conditions,
media composition

TISSUE CULTURE-
DERIVED PLANT

pathway of regeneration,
selection pressure,
genomic instabilities

PROGENY
ANALYSIS

However, somaclonal variation will find greater application for plant improvement in combination with selection for disease resistance, herbicide tolerance and other characters which are efficiently screened at the cellular level with *in vitro* cultures. Enhanced variation may contribute positively also in the context of interspecific hybridization. From the results described before, it is consequent to expect more genetic exchange between two genomes, which are combined sexually or asexually by protoplast fusion (Schieder 1982), when subsequently cultured for an extended period as *in vitro* cell culture. The obvious suppression or elimination of factors under cell culture conditions, which normally prevent the combination or exchange of genetic information in non-related germ line cells, may allow more recombination and thus enhance the possibility to integrate desired alien genes into crop plants more rapidly and efficiently than by conventional breeding methods.

ACKNOWLEDGEMENTS

I am indebted to Drs. W.R. Scowcroft and P.J. Larkin from C.I.S.R.O.
Canberra for many fruitful and encouraging discussions on this topic
and I want to thank Dr. P. Ozias-Akins for her help with the English
of the manuscript and Mrs. M. Pasemann for the typing.

REFERENCES

Barbier, M. and Dulieu, H.L. (1980). Effects génétiques observés sur
    des plantes de Tabac régéénérées à partir de cotylédons par
    culture *in vitro*. Ann. Amélior. Plantes 30, 321-344.

Bidney, D.L. and Shepard, J.F. (1981). Phenotypic variation in plants
    regenerated from protoplasts: the potato system. Biotechnol.
    and Bioengineering 23, 2691-2701.

Brettell, R.I.S., Thomas, E. and Ingram, D.S. (1980). Reversion of
    Texas male-sterile cytoplasm maize in culture to give fertile
    T-toxin resistant plants. Theor. Appl. Genet. 58, 55-58.

Browers, M.A. and Orton, T.I. (1982). A factorial study of chromosomal
    variability in callus cultures of celery (*Apium graveolens*)
    Plant Sci. Lett. 26, 65-73.

Burk, L.G. and Matzinger, D.F. (1976). Variation among anther-derived
    doubled haploids from an inbred line of tobacco. J. Hered.
    67, 381-384.

Carlson, P.S. (1974). Mitotic crossing-over in a higher plant. Genet.
    Res. Comb. 24, 109-112.

Chaleff, R.S. (1981). Genetics of Higher Plants. Applications of Cell
    Culture. Cambridge University Press, Cambridge.

Chourey, P.S. and Kemble, R.J. (1982). Transposition event in tissue
    cultured cells of maize, in Plant Tissue Culture 1982 (Fuji-
    wara, A., ed.) pp. 425-426, Jap. Assoc. of Plant Tissue Cul-
    ture, Maruzen, Tokyo.

D'Amato, F. (1977). Cytogenetics of differentiation in tissue and cell
    cultures, in Applied and fundamental aspects of plant cell
    tissue and organ culture (Reinert, J. and Bajaj, Y.P.S.,
    eds.) pp. 343-357, Springer Verlag, New York.

De Paepe, R., Prat, D. and Huguet, T. (1983). Heritable nuclear DNA
    changes in doubled haploid plants obtained by pollen culture
    of *Nicotiana sylvestris*. Plant Sci. Lett. 28, 11-28.

Edallo, S., Zucchinali, C., Perenzin, M. and Salamini, F. (1981).
    Chromosomal variation and frequency of spontaneous mutation
    associated with *in vitro* culture and plant regeneration in
    maize. Maydica 26, 39-56.

Fujiwara, A. (ed.) (1982). Plant Tissue Culture 1982. Jap. Assoc. of
    Plant Tissue Culture. Maruzen, Tokyo.

Gengenbach, B.G., Connelly, J.A., Pring, D.R. and Conde, M.F. (1981).
    Mitochondrial DNA variation in maize plants regenerated du-
    ring tissue culture selection. Theor. Appl. Genet. 59, 161-
    167.

Heinz, D.J., Krishnamurthi, M., Nikell, L.G. and Maretzki, A. (1977).
    Cell, tissue and organ culture in sugarcane improvement, in
    Applied and Fundamental Aspects of Plant Cell, Tissue and Or-
    gan Culture (Reinert, J. and Bajaj , Y.P.S., eds.) pp. 3-17,
    Springer Verlag, Berlin.

Javier, E.L., Burk, L.G. and Hanson, W.D. (1983). Variability among
    non-conventionally derived diploid lines of Nicotiana taba-
    cum. submitted.

Karp, A., Nelson, R.S., Thomas, E. and Bright, S.W.J. (1982). Chromo-
    some variation in protoplast-derived potato plants. Theor.
    Appl. Genet. 63, 265-272.

Kemble, R.J., Flavell, R.B. and Brettell, R.I.S. (1982). Mitochondrial
    DNA analysis of fertile and sterile maize plants derived from
    tissue culture with the Texas male sterile cytoplasm. Theor.
    Appl. Genet. 62, 213-217.

Larkin, P.J. and Scowcroft, W.R. (1981). Somaclonal variation - a no-
    vel source of variability from cell cultures for plant im-
    provement. Theor. Appl. Genet. 60, 197-214.

Larkin, P.J. and Scowcroft, W.R. (1983). Somaclonal variation and eye-
    spot toxin tolerance in sugarcane. Plant Cell, Tissue and
    Organ Culture, in press.

Lörz, H. and Scowcroft, W.R. (1983). Variability among plants and
    their progeny regenerated from protoplasts of Su/su hetero-
    zygotes of Nicotiana tabacum. Theor. Appl. Genet. in press.

Marx, J.L. (1983). A transposable element of maize emerges. Science
    219, 829-830.

McCoy, T.J., Phillips, R.L. and Rines, H.W. (1982). Cytogenetic ana-
    lysis of plants regenerated from oat (Avena sativa) tissue
    cultures; high frequency of partial chromosome loss. Can. J.
    Genet. Cytol. 24, 37-50.

McCoy, T.J. and Phillips, R.L. (1982). Chromosome stability in maize
    (Zea mays) tissue cultures and sectoring in some regenerated
    plants. Can. J. Genet. Cytol. 24, 559-565.

Menz, K.M. and Neumeyer, G.F. (1982). Evaluation of five emerging
    biotechnologies for maize. Bio Science 32, 675-676.

Ogihara, V. (1981). Tissue culture in *Haworthia*. Part 4: Genetic characterization of plants regenerated from callus. Theor. Appl. Genet. 60, 353-363.

Orton, T.J. (1980). Chromosomal variability in tissue culture and regenerated plants in *Hordeum*. Theor. Appl. Genet. 56, 101-112.

Prat, D. (1983). Genetic variability induced in *Nicotiana sylvestris* by protoplast culture. Theor. Appl. Genet. 64, 223-230.

Sacristan, M.D. and Melchers, G. (1969). The caryological analysis of plants regenerated from tumorous and other callus cultures of tobacco. Mol. Gen. Genet. 105, 317-333.

Schieder, O. (1982). Somatic hybridization: a new method for plant improvement, in Plant Improvement and Somatic Cell Genetics (Vasil, I.K., Scowcroft, W.R. and Frey, K.J., eds.) pp. 239-253, Academic Press, New York.

Schieder, O. (1983) Aktuelle Züchtungsforschung mit Arzneipflanzen: Ergebnisse und Perspektiven, in Biogene Arzneistoffe (Czygan, F.-C. ed.) F. Vieweg Verlag, Wiesbaden, in press.

Scowcroft, W.R. and Larkin, P.J. (1983a). Somaclonal variation, cell selection and genotype improvement. Comprehensive Biotechnology rd. 3, in press.

Scowcroft, W.R. and Larkin, P.J. (1983b). Somaclonal variation and genetic improvement of crop plants, in Better Crops for Food, Ciba Foundation Symposium No. 97, in press.

Secor, G. and Shepard, J.F. (1981). Variability of protoplast-derived potato clones. Crop. Sci. 21, 102-105.

Shepard, J.F., Bidney, D. and Shahin, E. (1980). Potato protoplasts in crop improvement. Science 208, 17-24.

Shepard, J.F. (1981). Protoplasts as sources of disease resistance in plants. Ann. Rev. Phytopath. 19, 145-166.

Skirvin, R.M. (1978). Natural and induced variation in tissue culture. Euphytica 27, 241-266.

Thomas, E., Bright, S.W.J., Franklin, J., Lancaster, V., Miflin, B.J. and Gibson, R. (1982). Variation amongst protoplast-derived potato plants (*Solanum tuberosum* c.v. 'Maris Bard'). Theor. Appl. Genet. 62, 65-68.

Vasil, V. and Vasil, I.K. (1981). Somatic embryogenesis and plant regeneration from suspension cultures of pearl millet (*Pennisetum americanum*). Ann. Bot. 47, 669-678.

Vasil, I.K., Vasil, V., Lu, C., Ozias-Akins, P., Haydu, Z. and Wang, D.Y. (1982). Somatic embryogenesis in cereals and grasses, in Variability in Plants Regenerated from Tissue Culture (Earle, D.E. and Demarly, Y. eds.) pp. 3-21, Praeger, New York.

Wenzel, G., Schieder, O., Przewoźny, T., Sopory, S.K. and Melchers, G. (1979). Comparison of single cell culture derived *Solanum tuberosum* L. plants and a model of their application in breeding programs. Theor. Appl. Genet. 55, 49-55.

DISCUSSION

P. STARLINGER: You mentioned that your variability is tissue culture-derived and that it occurred in the absence of any mutagenic treatment. Can you be sure that the 2,4-D has no mutagenic effects?

H. LÖRZ: We have to assume that tissue culture by itself is a somewhat mutagenic treatment for plant cells. The most obvious candidates acting as mutagens in the culture medium are the phytohormones, especially synthetic auxins. There is evidence from cytological analysis of long-term cell cultures that high levels of e.g. 2,4-D or NAA cause chromosomal abnormalities. Whether or not other media components (cytokinines, vitamins, sugars, salts, agar) have any "mutagenic" effect is not yet known.

STRATEGIES IN THE
IMPROVEMENT OF PLANTS

GENE TRANSFER IN MAIZE:
CONTROLLING ELEMENTS AND THE ALCOHOL DEHYDROGENASE GENES

W.J. Peacock, E.S. Dennis, W.L. Gerlach, D. Llewellyn, H. Lorz[1],
A.J. Pryor, M.M. Sachs, D. Schwartz[2], W.D. Sutton[3]

CSIRO Division of Plant Industry, P.O. Box 1600,
Canberra City, 2601, Australia

[1] Present address: Max Planck Institut fur Zuchtungsforschung,
D-5000 Koln 30, Federal Republic of Germany

[2] Present address: Department of Biology, Indiana University,
Bloomington, Indiana 47401, USA

[3] Present address: Plant Physiology Division, DSIR, Private Bag,
Palmerston North, New Zealand

INTRODUCTION

The world's principal crop and pasture species seem certain to provide
opportunities for gene surgery as an additional tool in plant
improvement. These few species have been well-tailored genetically,
yet they are frequently challenged by pests and diseases, and by other
factors which result from changing agronomic conditions. Many of the
challenges have been met by plant breeders introducing new single
genes; a fact which is important if recombinant DNA technology is to be
used in plant improvement since a molecular biologist will be
restricted for some time to relatively simple genetic manipulations.
At present when a plant breeder uses traditional methods to meet the
challenge he cannot avoid introducing approximately 50,000 gene
equivalents together with the gene he actually needs to overcome the
defficiency in the crop species. He then has the demanding and time-
consuming task of eliminating many of these superflous genes and
recovering an approximation to the well-tailored genome which
previously existed for that crop species but which in addition now
carries the required gene. Gene surgery or gene transfer by genetic
engineering may allow the introduction of a single required gene into
an otherwise satisfactory genotype without causing the large-scale
perturbations associated with traditional breeding processes.

The challenges posed by new diseases to well-adapted agricultural species can be of significant dimensions. It is probable that diseases alone account for a 15-25% loss in potential yields even in well-developed agricultural systems. If resistance to these diseases could be readily introduced into our existing high-performance genotypes the returns would be large. For example, the accidental introduction of two aphid species resulted in the devastation of lucerne (Medicago sativa) in Australia in the late 1970s. The lucerne industry was largely based on a monoculture with the cultivar Hunter River and this was highly susceptible to the aphid. This particular challenge was met in Australia by a number of plant breeding programs which have now produced aphid resistant varieties (Oram, 1980). Although now available at a commercial level, these cultivars are still not ideally adjusted to Australian requirements. In breeding programs which used hybridization to cultivars from the United States which carried aphid resistance genes, back-crossing to traditional Australian cultivars is still needed to produce an ideal variety for Australian conditions. It would have been of great value if resistance genes could have been isolated and cloned from the American lines and introduced to the existing susceptible Australian lines without large-scale disturbance of the genotype.

Even in the most highly-developed agricultural species such as hexaploid bread wheat (Triticum aestivum) there will remain significant opportunities for single gene characters to produce an advantage in yield. Significant yield increases followed the incorporation of the gene Rht 1 into Australian cultivars in the last decade (Pugsley, 1983).

Another likely use of recombinant DNA technology in plant improvement is where required genes must be introduced from another species not able to hybridize with the agricultural species. An example is the Australian native species Glycine canescens, related to soybean, which contains a gene for resistance to soybean rust (Phakopsora pachyrhizi) (Burdon and Marshall, 1981). This rust is of major significance in South East Asia where probably 30% of potential yield is lost almost every year and where the soybean crop is devastated over large areas. The rust is already in Australia and can be expected to affect other soybean regions of the world. Experiments in the Division of Plant Industry have shown that the gene from native Australian species does confer resistance in an Fl hybrid with soybean. A substantial and difficult breeding program is required before the Fl hybrid can be integrated back into a suitable commercial soybean cultivar background. Again, if the single resistance gene could be identified and isolated from the native Glycine species and integrated into the soybean genome by recombinant DNA techniques, it would be of very considerable advantage.

RECEPTOR CELL SYSTEMS

The paper by J. Schell in this Symposium has reported the advances in the Agrobacterium T-DNA system which have resulted in the introduction of bacterial genes in working order into a plant genome. This system, based on the crown gall organism, is not available for many important agricultural crops which are monocots. In our Laboratory we are attempting to develop a gene transfer system for maize which is a monocot, and our hope is that the system may be of general use in plants. Its two principal components are the alcohol dehydrogenase genes of the maize genome (Schwartz, 1966; Freeling and Schwartz, 1973), and McClintock's Ac/Ds controlling elements (McClintock, 1951). There are mutants of Adh1 induced by the Ds controlling element (Osterman and Schwartz, 1981) and we reasoned that cloning of the Adh1 gene would enable subsequent cloning of the Ds controlling element.

Maize seedlings require ADH enzyme activity to survive anaerobic conditions (Schwartz, 1969), and we considered that this may equally apply to maize cells in culture. If this proved to be so then cells derived from a line known genetically to be Adh11$^-$ and Adh2$^-$ would not survive under anaerobic conditions nless a working Adh$^-$ gene was introduced. We have shown that anaerobiosis, induced by flooding the cells with argon, does provide an effective selection against cells lacking a working Adh gene (Figure 1). With cell cultures derived from fifteen-day embryos the length of argon treatment provides an effective screening in favour of Adh$^+$ and against Adh$^-$ cells. Aggregates of Adh$^+$ cells will regenerate plants following the removal of the anaerobic conditions, but Adh$^-$ cells will not.

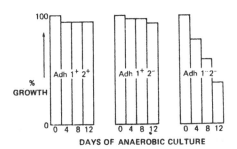

DAYS OF ANAEROBIC CULTURE

Figure 1. Comparison of growth of callus cells derived from scutellum of immature maize embryos of different alcohol dehydrogenase genotypes when subjected to anaerobic conditions. The graph shows callus weight (as a proportion of control aerobic callus) after anaerobic treatments of 0,4,8,12 days.

These results give promise that if the Adh$^+$ genes can be introduced into Adh$^-$ cells they will provide a selection system for gene transformation.

However, in maize we have no cell culture system which enables us to produce protoplasts and regenerate plants from them. Protoplasts are desirable because they greatly simplify the efficient introduction of DNA. For this reason we have begun working with Nicotiana plumbaginigolia, a species which provides an efficient protoplast isolation and subsequent plant regeneration system. The difficulty with N. plumbaginifolia is that there are no Adh$^-$ genotypes available. We have treated N. plumbaginifolia protoplasts with a mutagen and have

selected for <u>Adh</u> mutants by the allyl alcohol screening procedure
(Schwartz and Osterman, 1976).   In the presence of ADH enzyme allyl
alcohol is converted to the toxic aldehyde, acrolein, and the cells
die.   In the absence of ADH enzyme the cells survive the allyl alcohol
treatment.   Allyl alcohol resistant colonies have been retested and
regenerated into plants, which now need to be further tested to
determine if the <u>Adh</u>⁻ condition is heritable.   And <u>Adh</u>⁻ N.
<u>plumbaginifolia</u> should provide us with the opportunity of using a
protoplast-regenerative system for <u>Adh</u> gene transfer.

Another approach we are using is to incorporate an <u>Adh</u> gene from maize
into the nopaline synthase gene in the T-DNA region of the plasmid of
the crown gall organism <u>Agrobacterium tumefasciens</u> (Leemans et al.,
1981).   We have introduced the <u>Adhl</u> gene and its promoter region into
the <u>Agrobacterium</u> plasmid (Figure 2) and have produced galls on the
stem tissue of <u>Arabidopsis thaliana</u> using a genotype lacking an ADH
enzyme (Jacobs and Dolferus, 1983).

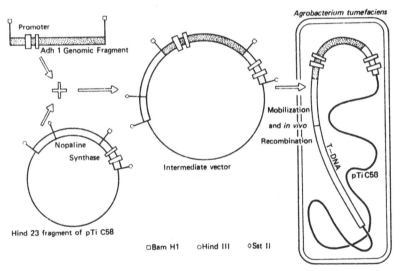

Figure 2.   Insertion of <u>Adhl</u> gene into the nopaline synthase
gene of the <u>Agrobacterium tumefasciens</u> Ti plasmid.

THE MAIZE <u>Adh</u> GENE SYSTEM

There are two <u>Adh</u> genes in maize, <u>Adhl</u> on chromosome 1 and <u>Adh2</u> on
chromosome 4.   Both ADH polypeptides are about 40,000 daltons and <u>in</u>
<u>vitro</u> they form a functional heterodimer, suggesting that the two
genes are probably derived from an ancient duplication.   In order to
use one or both of these genes in gene transfer experiments we isolated
both cDNA (Gerlach et al., 1982) and genomic versions of the genes.   We
sequenced the genes and their flanking regions in order to pinpoint the
extent of the coding regions and the positions of the transcription and
translation controls.

We assumed that anaerobic induction of the enzyme resulted from anaerobic induction of transcription and we made cDNA libraries from mRNA isolated from anaerobically treated maize seedlings (Jacobs and Dolferus, 1983). We identified those cDNA inserts which showed up positive with anaerobic mRNA probe and negative with aerobic mRNA probe, and by a number of techniques were able to identify those which were derived from the Adh1 gene transcript. The cDNA segment was then used as a probe for the Adh1 genomic sequences. These were isolated from libraries of clones generated in λ1059 using either BamHl or Sau3A digests of maize DNA cloned into the Bam site. A genomic Adh1 segment has been mapped with restriction enzymes and sequenced (Figure 3).

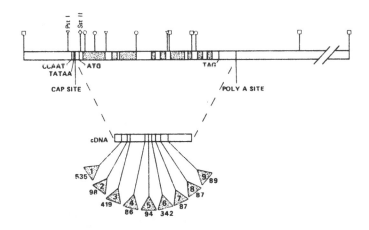

Figure 3. Structure of the Adh1-1S gene including intervening sequences. Introns are indicated as shaded areas and their lengths in base pairs are shown in the lower part of the diagram. Some restriction enzyme sites are shown: Squares Bam Hl, inverted triangles Pst1, diamonds Sst2, open circles HindIII, closed circles Bgl2, closed squares Sal1, closed inverted triangles are Pvul. ATG and TAG are transcription start and end points respectively.

We have also isolated Adh2 cDNA and genomic segments but in this paper we will confine our remarks to the Adh1 gene. It contains nine introns which obey the usual eukaryotic rules with regard to consensus junction sites. We have used nuclease S1 mapping to identify the transcription start site and sequencing has established the positions of the probable TATAA and CCAAT boxes (Figure 4). As yet we have not identified any sequence in the 5' region responsible for the anaerobic induction of the Adh1 and Adh2 genes. However, we have identified the principal land marks of the gene which are of importance in making constructs to be used in expression and transformation studies.

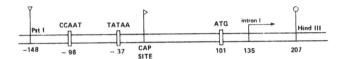

Figure 4. 5' region of the Adhl gene. The positions of the presumptive CCAAT and TATAA boxes, the transcription initiation site (CAP) and translation initiation site (ATG) are shown along with the position of the first intervening sequence of the gene. The numbers refer to base pairs distant from the transcription initiation site.

We have confirmed that the anaerobic induction of Adhl is at the transcriptional level and have determined the length of mRNA for Adhl to be approximately 1650 bases. This applies to a particular electrophoretic allele which we have used for most of our studies. In some other alleles there is an additional messenger of length approximately 1750 bases. The molecular basis of the additional mRNA species has not yet been determined.

THE Ds CONTROLLING ELEMENT

Our interest in the mobile controlling elements in maize is two-fold. We assume that they may prove to be useful as a component of a vector to introduce genes into maize chromosomes in the same way that the P elements have been used in Drosophila melanogaster (Rubin, 1983). Our second interest was that these mobile DNA segments may be useful as transposon mutagens in maize, allowing the identification of genes of agricultural importance.

Osterman and Schwartz (1969) isolated a mutant of Adhl apparently resulting from an insertion of the Ds controlling element. The mutant gene produced only one-tenth of the normal amount of the enzyme activity and it also had an increased thermolability. The mutant gene reverted at high frequency to a gene producing an enzyme with normal levels of activity in the presence of Ac, the autonomous element of the Ac/Ds system (McClintock, 1951). Osterman and Schwartz recovered some of these Ac-induced revertants. They have provided us with (a) the progenitor (PR) stock in which Ds was known to be inserted at the bz2 locus in the distal region of the long arm of chromosone 1 (as is Adhl), (b) the Ds stock (DS) in which Ds had been introduced into the Adhl gene by Ac, and (c) revertant stocks (RV) in which Ac action has apparently removed the Ds element from the Adhl gene restoring normal gene activity (Figure 5).

We isolated the Adhl gene segments from these stocks and mapped them using restriction enzymes (Figure 6). The Adhl gene appeared to be identical in all three stocks except for one restriction fragment, Pstl-Sst2, which we knew to be in the 5' region of the gene and which contained the transcription and translation initiation control sequences. The resctriction map of this region suggested there was approximately a 400 bp increased length of the region in the Ds mutant, consistent with an approximately 400 bp insertion.

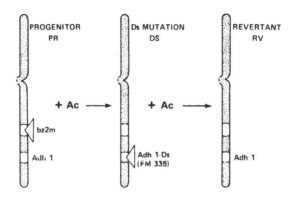

Figure 5. Maize stocks involving the <u>Adhl</u> locus and the Ds controlling element in maize, after Osterman and Schwartz (1981).

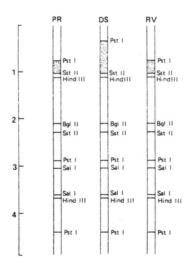

Figure 6. Restriction enzyme mapping of the progenitor (PR), Ds mutant (DS) and revertant (RV) <u>Adhl</u> gene regions.   The scale is kilobase pairs.

Nucleotide sequencing of this 5' zone established that in the Ds mutant there was a 405 bp insertion between nucleotides 45 and 46 of the transcribed region, thus placing the Ds insertion between the transcription and translation start points (Figure 7).   The analysis

showed the boundaries of the insertion to be an inverted repeat of 11
bp. It is conceivable that this terminus is slightly more complex
since two bases along from the 11 bp repeat there is a 3 bp segment
which is in a comparable inverted condition. We have sequenced the
entire insert and have not detected any more extended inverted
repeats. The central region of this Ds insertion is extremely AT-rich
(76%).

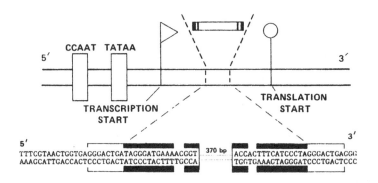

Figure 7. Ds insertion into Adh1 gene. The insertion is
between nucleotides 45 and 46 of the transcribed region and is bounded
by 11 bp inverted repeats (shaded).

The sequence also showed that the insertion of the Ds element generated
a direct duplication of an 8 bp genomic segment. The Ds element thus
has the properties of inverted terminal repeats and induced genomic
direct repeats which have been described for a number of other
eukaryotic and prokaryotic insertion elements.

The revertant stock does not completely return to the progenitor
condition. In this stock the Ds element had been excised but in a
manner which left the 8 bp duplication in the genome (Figure 8). The
first and second members of the now tandem 8 bp duplication contained
mutated nucleotides in the revertant. The 3' nucleotide of the first
of the 8 bp segments was altered and the most 5' nucleotide of the
second element of the tandem duplication was also altered. Presumably
these changes are in the bases immediately adjacent to the site of
excision of the Ds element. Sequencing of a second independent
revertant established that these same two nucleotide substitutions had
occurred but in addition the second nucleotide of the second element in
the duplication was also mutated. In each of these cases the mutation
is to the base complementary to the pre-existing base and presumably
this reflects a well-defined excision mechanism. The mechanism of
excision must not involve a recombinant event between the genomic
direct repeats since this would reduce the duplication back to a
single representation of the 8 bp segment.

One intriguing feature of this Ds mutant is that we have shown the insertion to be 5' of the coding region, yet Osterman and Schwartz (1981) determined that the Adh1 enzyme itself was more thermolabile than the normal enzyme. We therefore suppose that the Ds element must in some way contribute to a modified protein. Since in eukaryotes generally the most 5' ATC codon is the operative translation initiation codon, we asked whether there were any ATG codons in the Ds element. There are six ATGs in the Ds element but all of them are followed after relatively few codons by a stop codon. It seems therefore that translation does not begin in the Ds element to produce a much larger fusion enzyme. This is in agreement with the data of Osterman and Schwartz who demonstrated that the mutant protein was of size comparable to that of the normal protein. Furthermore our own data (unpublished) suggests that the mRNA is of comparable size to the normal mRNA. However, initiation of translation could still begin at the first ATG in Ds provided that there is an excision of an intron segment such that the mutant mRNA is still comparable in size to the normal mRNA. Detailed sequence analysis has shown that such a situation is possible. An acceptable intron donor sequence, in the

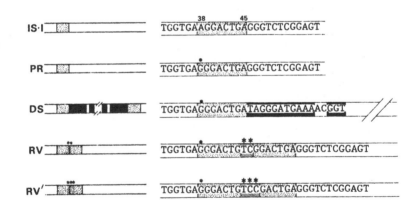

Figure 8.   Structural and sequence characteristics of the progenitor, Ds mutant and revertant stocks. 1S.1 is the sequence from the standard Slow allele of Adh1. The progenitor (PR), Ds mutant (DS), and two independent revertants (RV and RV') differ from the standard Slow allele in nucleotide 38 (small asterisk). The mutated nucleotides in the revertants are denoted by the large asterisk.

same reading frame as the first intron donor sequence of the normal Adh1 gene, occurs 10 codons after the first ATG codon which is within the inverted repeat of the Ds element (Figure 9). The proposed scheme substitutes 10 amino acids coded from this portion of the Ds sequence for the 11 amino acids normally found in the Adh1 enzyme. The changes are such that they could be compatible with a changed thermolability of the molecule. This postulate is being tested by sequencing the S_1 mapping of the 5' region of the Ds mutant gene sequence.

When we used the cloned Ds insertion to probe restriction-digested DNA of maize we found that in all stocks we examined there were 30-40 positive signals ranging in size from 30 kb down to approximately 0.3 kb. This suggests that there are a number of Ds elements normally present in the maize genome. Cloning and sequencing of a number of

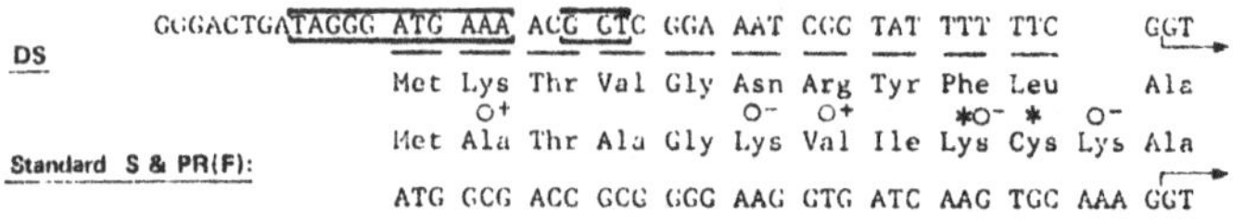

Figure 9.   Possible N-terminus in ADH of the Ds mutant. Asterisks mark amino acid substitutions which could be significant for properties of the protein. Other charge changes are indicated. Arrows indicate the beginning of intron no. 1.

these apparent Ds elements will be necessary to establish whether they all have the same termini with the 11 bp inverted repeat. It will also provide evidence as to whether any sequence homology exists between the point of insertion of the Ds element in the genome and the terminal sequence of the Ds element. Homology is suggested by the one case we have examined in detail (Figure 10). If homology is required, as seems to be the case in TN10 in Escherichia coli (Halling and Kleckner, 1982), this would have implications as to the number of potential sites in the genome into which Ds might insert either as a vector or as a mutating transposon.

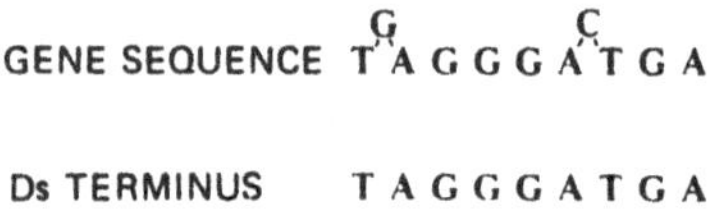

Figure 10.   Homology of the Ds insertion site in Adhl gene with that of the terminus of the Ds element.

GENE TRANSFER IN MAIZE

Any use of Ds as a vector for gene transfer of maize will have to be done in the presence of the Ac element. We are currently attempting to

isolate Ac but have also approached the problem by making maize receptor cells from genetic stocks which do include Ac in their genotype. McClintock (1951) established that the Ac/Ds system is operative in somatic as well as germline cells so it seems reasonable to assume that it will be operative in cultured maize cells.

If the Ac/Ds system does not turn out to be comparable to the P element system in Drosophila melanogaster then it should prove to be a valuable component of a gene transfer system. The Ds jumping gene could for example be used to transfer a disease-resistance gene into a susceptible agricultural stock.

We have not yet demonstrated gene transfer in maize but we now have isolated what we believe to be the most important components of such a system. The Ds element could provide an efficient mechanism of gene insertion, the <u>Adhl</u> gene should enable an efficient selection of transformed cells to be made, and one or other of the receptor cell systems we have been developing may enable us to regenerate $\underline{Adh}^+$ plants from the selected transformed $\underline{Adh}^+$ cells.

ACKNOWLEDGEMENTS

We are grateful to K. Ferguson, Y. Hort, M. Jeppesen, G. Koci, J. Norman and A. Tassie for their skilled assistance in this work.

REFERENCES

Burdon, J.J. and Marshall, D.R. (1981). Plant Disease, 65. 44.

Freeling, M. and Schwartz, D. (1973). Biochem. Genet. 8, 27.

Gerlach, W.L., Pryor, A.J., Dennis, E.S., Ferl, R.J., Sachs, M.M. and Peacock, W.J. (1982). Proc. Natl. Acad. Sci. USA 79, 2981.

Halling, S.M. and Kleckner, N. (1982). Cell 28, 155.

Jacobs, M. and Dolferus, R. (1983). "Advances in Gene Technology: Genetics of Plants and Animals", Miami Winter Symposium, Vol. 20. Academic Press, New York.

Leemans, J., Shaw, C., DeBlaere, R., DeGreve, H., Hernalstens, J., Van Montagu, M. and Schell, J. (1981). J. Mol. Appl. Genet. 1, 149.

McClintock, B. (1951). Cold Spring Harb. Symp. Quant. Biol., Cold Spring Harbor, New York.

Oram, R.N. (1980). J. Austral. Inst. Agric. Sci. 46, 200.

Osterman, J.C. and Schwartz, D. (1981). Genetics 99, 267.

Pugsley, A.T., (1983). Plant Disease 65, 44.

Rubin, G.J. (1983). "Advances in Gene Technology: Molecular Genetics of Plants and Animals", Miami Winter Symposium, Vol. 20. Academic Press, New York.

Schwartz, D. (1966). Proc. Natl. Acad. Sci. USA 56, 1431.

Schwartz, D. (1969). Am. Nat. 103, 479.

Schwartz, D. and Osterman, J. (1976). Genetics 83, 63.

DISCUSSION

P. STARLINGER: How sure can you be that the Ds in Fm335 was transposed
from bz2-m? The Ds in 2F" isolated by Doring came from the same bz2
background, but differs in size from Fm335. Would you consider the
possibility that one or both came from one of the other 40 sites that
we have shown to hybridize to Ds?

W.J. PEACOCK: Until we have more knowledge of the other apparent Ds
elements in the genome, I do not think I can give a useful answer. If
we examine directly the Ds element at the bz2 locus, then we would have
the answer.

E. RUDIGER: How are introns distributed with respect to NAD- and
substrate-binding domains? Are introns located at identical positions
in Adh1 and Adh2 genes?

W.J. PEACOCK: There is 50% absolute homology between the amino-acid
sequence of horse-liver alcohol dehydrogenase and the sequence
predicted from our nucleotide sequence analysis and the maize Adh
gene. The horse-liver enzyme has two distinct functional components,
a catalytic region and a more C-terminal coenzyme-binding region. The
probable boundary of similar regions in the maize enzyme coincides
closely with intron and exon junctions such that only exons 5, 6, 7, 8
and 9 comprise the coenzyme- or NAD-binding region. At present the
knowledge of further functional domains does not permit us to say
anything with regard to their correspondence with intron/exon
positions. However, there are two regions of high conservation of
amino-acid sequence at the N-terminus, and the first of these regions
bridges exon 1 and exon 2, suggesting a lack of correlation between the
intron position and the possible functional domains.

In answer to the second question, we still do not have complete
knowledge on precise intron positions in the Adh2 gene but they are
obviously extremely similar, if not identical, to those in the Adh1
gene.

POTATO - A FIRST CROP IMPROVED BY THE APPLICATION OF MICRO-
BIOLOGICAL TECHNIQUES ?

G. Wenzel[1], H. Uhrig[2], and W. Burgermeister[3]

Federal Biological Research Centre for Agriculture and
Forestry
[1]Institute for Resistance Genetics, Grünbach, F.R.G.
[3]Institute for Biochemistry, Braunschweig, F.R.G.
[2]Max-Planck-Institute for Plant Breeding Research, Cologne,
F.R.G.

ABSTRACT

Todate there exist just a few important crop plants, where cell and
tissue culture techniques are developed so far that an application in
the field becomes reasonable. Potato, *Solanum tuberosum*, is one of
them; here *in vitro* propagation, haploid induction at the di- and mo-
nohaploid level and protoplast techniques work, at least for several
genotypes. The combined application of these techniques together with
classical breeding procedures is described. Further, problems like
somaclonal variation and protoclone formation are discussed under an
applied aspect.

INTRODUCTION

The progress in understanding the basic genetics, beginning with microorganisms and today increasingly transgressing to higher plants,
opens up a chance of counterbalancing the contingency of evolution by
purposeful application of microbiological techniques in crop plant
breeding. Such techniques, summarized under the phrase unconventional
breeding procedures, are in particular: *in vitro* propagation, *in vitro*
selection, production of haploids, somatic hybridization, and gene
transfer. According to methods, the step from bacteria to higher
plants has been done already, but there is a tremendous gap between
results obtained with a few model plants like tobacco and *Petunia*, and
field experiments. The question of when these techniques finally will
pay in the field cannot be answered yet. One of the most attractive
objects for clarification is potato. It is 1. in the *Solanaceae*, the
family most model plants belong to; it is 2. a vegetatively propagated
crop, permitting the maintenance of new genotypes for generations and
3. it is a plant of strong economic interest and a steadily expanding
growing area, at least in the subtropics.

One of the strongest obstacles against efficient breeding work is the
tetraploid nature of potato. Today a breeder is forced to start with
5oo,ooo - 1,ooo,ooo seedlings in order to bring one new variety to the
market. The breeder produces such F_1-populations by combination bree-

ding and the classical breeding approach is exclusively selection,
with hardly any possibility of influencing the rate of success. The
whole process needs 8 - 15 years. Further,there is a great danger of
discarding good material during early screening. Breeding at lower
ploidy levels using dihaploids with 2n = 2x = 24 chromosomes and mo-
nohaploids with 2n = x = 12, is consequently a central step in unconventional potato breeding, permitting at least a more directed selection in smaller populations. The present selection rates for haploids
are given in Tab 1. Before discussing the use of haploids, we would
like to mention results obtained using vegetative propagation in potato improvement.

TABLE 1. Efficiency of different procedures for haploid induction in *Solanum tuberosum*.

2x	4x	Parthenogenesis number of				Androgenesis number of	
		fertilized flowers (1oo%)	seeds	haploid plants	anthers (%) plated	plants regenerated	(%)
	2x	3,938	6,854	1,824	46 2,700	16	o.6
1x		7,965	262,648	5	o.6 54,931	2,317	4

RESULTS

Vegetative propagation. One of the first applications of tissue culture
in potato was the use of meristem culture for propagation of specific
genotypes. This method only gained credit, however, when coupled to
another, normally a phytosanitary aim. It should be emphasized here
that ten-thousands of tubers have been produced during the last decade
via meristems, being altered neither in geno- nor in phenotype. During
1977 one of the leading German potato varieties, 'Hansa' was completely
infected by a new strain of PVY; it was cured from the virus by meristem culture within two years, and all 'Hansa' grown today, descends
from meristems - and 'Hansa' is still 'Hansa'. One reason for this
stability is surely the fact that during meristem culture no strong
dedifferentiation takes place, although via cytokinin action multiple
shoot formation with some callus formation is induced. A stronger dedifferentiation happens in procedures with prolonged callus cultures,
such as *in vitro* selection, which then may cause enhanced phenotypic
and/or genotypic variability.

In vitro selection. In an experimental series started by Behnke (1979)
we focussed on disease resistance using two fungi, *Phytophthora infestans* and *Fusarium ssp.*, which damage the potato by exotoxins. Both
exotoxins are excreated into liquid fungal growth medium. In the presence of crude filter extracts of the fungal growth media, some calli
survived, and in the case of *Phytophthora*, plants could be regenerated.
The results of such an experiment are given in Fig. 1. As the starting
material was selected only for a high regeneration capacity, and not
for other valuable characters, the majority of the regenerants was lost
in the field because of other diseases.

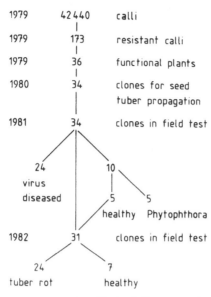

Figure 1. Selection for *Phytophthora* resistant potato clones starting from dihaploid calli.

In order to get some idea of what might have happened during or before selection, we tested the protein patterns of selected resistant and susceptible clones and compared them with the starting material, clone HH 345, an interdihaploid *Solanum tuberosum* clone. The pherogram on polyacrylamid gels run in Tris-boric acid buffer at pH 7.9, 16omin for electrophoresis and in a pH 4 - 9 gradient, 2,6oo Volt hours for electrofocussing showed only weak differences (Fig. 2a, b). It was not possible, to correlate one specific alteration to a specific behaviour during selection.

The resistance has been maintained over several years of vegetative propagation, which is a strong argument for a stable mutation. On the other hand the pherogram patterns of the total proteins revealed no difference between resistant and susceptible clones and additionally the number of resistant calli could not be increased significantly by mutation using 2 Kr for callus irradiation (Behnke 1979). Although such a behaviour points in the direction of epigenetic effects, it can be explained by a preexistance of a high number of mutations covering additional ones. The available data make it probable that indeed mutations have taken place. However, spontaneous or tissue culture induced mutants will not improve potato breeding generally. It has been shown convincingly that purposeful approaches using combination breeding are better than just raising the variability of a basis population. This variability may only gain some credit, it in tissue culture specific mutations appear or if it can be combined with a powerful selection system.

Protoplasts. By reducing the size of the explants from callus to cells, one reaches the protoplast level. In potato protoplast regeneration is not only possible at the tetraploid, but also on the dihaploid level.

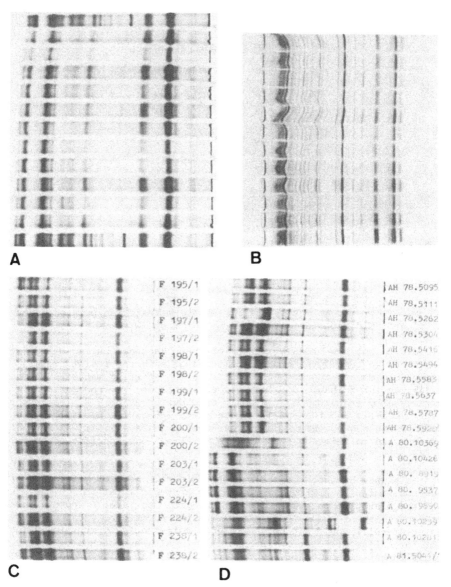

Figure 2. Tuber protein patterns of *Solanum tuberosum*. A) Pherogram of clones regenerated from the dihaploid clone HH 345 (C = control)
and clones regenerated after *in vitro* selection on medium containing
exotoxin of *Phytophthora infestans* (R = resistant in field test; S = susceptible). B) The same material in electrofocussing. C) Patterns of
clones regenerated from protoplasts of the dihaploid clone HH 260. D)
Protein patterns of clones regenerated from microspores of dihaploid
potato clones. Identical pherograms make it feasible that secondary
cloning happened, while differing patterns are indicative for the origin from different microspore genotypes.

In contrast to meristem culture, the protoplast regeneration passes always a phase of dedifferentiation, followed by unorganized callus growth and finally organogenesis. It is under discussion, how far this procedure beneficially alters the genotype. Shepard (1982) found such a tremendous variability in clones regenerated from protoplasts of tetraploid cultivars, especially from 'Russet Burbank' that he recommended the use of such protoclones for intracultivar improvement. We regenerated more than 3,000 clones from protoplasts of dihaploid potatoes at the Max-Planck-Institute; we found striking aberrants (Fig. 3), most of which could be identified as aneuploids. (A job, which is not easy because of the very tiny potato chromosomes.) On the other hand, aneuploids are rather common and also stable in potato; a high proportion of them is even vigorous. In a cross e.g. of 3x X 2x they can be produced in large numbers (Stabel 1982; Fig. 4). Having discarded such distinct aberrants, the rest of the protoplast-derived plants looked phenotypically rather uniform - at least we could not detect stronger differences than in a normal potato clone growing in the field. An aliquot of 36 protoplast-derived clones, most of which had doubled to the tetraploid level during *in vitro* culture, were screened by polyacrylamide gel-electrophoresis under the conditions described above. As figure 2c documents, there is hardly any difference in the total protein pattern after Coomassie blue staining. According to our classification system (Stegemann and Loeschcke 1976) the samples would be considered identical.

Figure 3. Phenotypic variation and abnormalities in protoplast derived potato clones.

G. Wenzel et al.

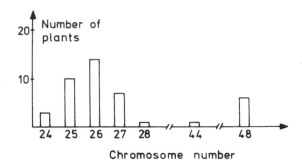

Chromosome number

Figure 4: Frequencies of aneuploid chromosome numbers in potato after 3x X 2x crossing (Stabel 1982).

We tried to keep the callus phase in protoplast regeneration as short
as possible, because genome changes are desired neither for protoplast
fusion nor for rapid propagation.

Use of haploids. The chance to improve potato by the use of protoplasts
and/or callus selection is predominantly opened up by the large number
of individual genotypes which can be handeled on a small area, the
Petri dish. It is not too critical if the *in vitro* culture *per se* creates new variability or not; what is important, however, is that a
powerful selection system is available. As we do not see at the moment
a chance to use accidental new variability for potato improvement, we
prefer an approach which is based upon the parental genes. We think
that the stepwise selection of desired genotypes at the different
ploidy levels and a logical reconstruction of tetraploids is highly
reproducible and will be finally more efficient (Wenzel et al. 1982).
The use of dihaploids - the first step in this process - is already
successful in the field.

In the combined application of parthenogenesis, androgenesis, protoplast isolation and somatic fusion, each single process works, but
there are some problems with identification of somatic hybrids. We
face, however, some additional problems as the success rate is highly
dependent on the genotype of the starting material. Some genotypes
respond actively in tissue culture, which means, they possess a good
regeneration capacity. As most of these clones do not contain valuable
agronomic characters, one has to produce firstly hybrids among clones
with a good tissue culture ability and clones carrying e.g. resistances. The sexual F_1 hybrids can be screened for the presence of the
resistance and then used as anther donor plants. In such hybrid anther
donor plants we found clear segregation for tissue culture ability,
and as we worked on the diploid level, the segregation rate gave us
some hints that this character is heritable (Uhrig 1983). As the results are in full agreement with the findings of genetically based
tissue culture ability in rape-seed (Hoffmann et al. 1982), rye (Wenzel et al. 1977) and barley (Foroughi-Wehr et al. 1982), we believe
that it is an universal situation. It is surely more expensive to develop a suitable medium for a recalcitrant genotype than to broaden the
genetic basis for tissue culture ability via selection and combination
breeding. But there exists the danger that we select in the segrega-

ting microspore population only those genotypes with a good tissue culture ability, especially since to date the number of regenerated plants
from one given donor type is never large enough to guarantee the sta-
tistical presence of most of the possible genotypes. There is no
question that selection takes place. It is, however, probable, according to the available results, that the selected types are at the same
time the most vigorous ones. This is congruent with the wishes of the
practical breeder. Additionally the passage via haploids purifies genomes from lethal alleles which are only sublethal in the diploid
state. By this purification the inbreeding depression is drastically
reduced.

Characters of androgenetic clones. A first point, important under the
applied aspect, is the fact that plants passing from 4x to 1x and back
to 4x still contain valuable characters, e.g. resistances, regardless
of their mode of inheritance (Wenzel and Uhrig 1981). We could detect
monogenic characters such as extreme virus resistance as well as polygenic ones such as field resistances. This is of importance, as there
is a strong tendency in breeding work, to base e.g. resistances no
longer on the small vertical but on the broader horizontal basis. Mo-
nogenically coded qualitative resistances are broken down from altered
pathotypes of the pathogen normally within 5 years, while quantitative
resistances last much longer. The importance of this finding is under-
lined by the situation of classical breeding: transfer of a monogenic
trait is no severe problem, but the breeders have no breeding scheme
ready for the combination of quantitative characters with reasonable
population sizes.

Protoplast fusion. For the combination of quantitative characters also
somatic hybridization is a challange. Fusion of two dihaploid potatoes
being heterozygous for polygenically inherited characters without any
meiotic mixing is extremely promising for the quantitative approach.
Consequently this is the final step in an analytical synthetic breeding scheme (Wenzel et al. 1982). Although the fusion itself is no problem, the selection of fusion products turned out to be difficult.
In the successful fusion experiments published, e.g. potato (x) tomato
(Melchers et al. 1978) or potato (x) *Atropa* (Binding et al. 1982), the
identification of fusion products was finally possible by the phenotype. Under applied aspects the fusion partners will be nearly identical and their selection via the phenotype is hopeless. To increase
the success rate in *S. tuberosum* (x) *S. tuberosum* somatic hybrid
identification, Uhrig (1980) developed a reversible bleaching system.
He could demonstrate that the herbicide SAN 6706 bleaches potato shoots
without blocking the regeneration capacity of bleached protoplasts.
When only one fusion partner was bleached, fusion products could be
identified , collected with a micropipette (Hein et al. 1983) and
cultured, either pooled or with a nurse culture.

Present problems in applying unconventional breeding steps in potato.
Besides the fact that no clear fusion products are available yet, we
found unexpected results in androgenetic clones in the field. During
the last years we selfed A_1 plants to obtain A_2's, and we produced
several hybrids between A_1's. In contrast to the expectation most A_2's
segregated and the F_1's were heterogeneous. There exist several possi-

ble explanations, the simplest one of which is that the heterozygous
A_1's were derived from unreduced diploid microspores. As the A_1's are
not identical, phenotypically rather different from the anther[1] donor
clones and additionally different amongst each other, one has to postulate further a high cross over rate, or a high mutability. Figure 2 d
shows that the pherograms of total protein of such plants contain
classes. Identical patterns of different clones show that these clones
descend from the same microspore and were propagated during early
microcallus stage. All other clones differ significantly in their
patterns, which is in agreement with the phenotypic observations. The
explanation of Straub (1973) for such variability observed in haploid
progenies of *Petunia* cannot be used here. He discussed that the haploid situation *per se* is not stable and creates new variability because,
within haploids, pairing of non-homologous chromosomes may take place.
After chromosome doubling in *Petunia* the genome has been stable again.
We found such a situation in doubled haploids of rape-seed (Hoffmann
et al. 1982) but the situation in potato needs a different explanation. As we further never observed such alterations in barley (Foroughi-Wehr et al. 1982), it is possible that such a variability is
more common in outbreeders. De Paepe et al. (1981) proposed as an explanation for similar phenomena in *Nicotiana sylvestris* that the genotype of the vegetative and the generative nuclei is different because of different amounts of endoreduplication. By fusion of such

Figure 5. Hybrid vigor (middle) in a potato clone combined from
two A_1 clones (right and left).

different nuclei, microspores arise with genomes heterozygous for some chromosome regions. Brown and Wernsmann (1982) report on similar heterogeneous androgenetic populations in tobacco, coupled to a productivity depression, which was even not dissipated with the establishment of new heterozygosity. In potato we clearly found hybrid vigor after crossing androgenetic clones regardless if these were homo- or heterogeneous (Fig. 5). This heterosis was expressed in green plants as well as in the tubers.

CONCLUSION

During *in vitro* culture of potato, two phenomena could be observed: 1. Uniformity of regenerants as long as the callus phase was kept short and 2. increasing mutation (?) rates with increasing *in vitro* periods, creating new variability, which might perhaps be useful for *in vitro* selection. In addition microspore-derived clones showed variability, the reason for which is still open. As long as we do not know the actual reason for the new variability, we should be careful in praising it as a new breeding tool. The failure of mutation breeding in higher plants teaches that a procedure which just creates new variability is not superior to classical combination breeding, as long as genetic resources are available in a compatible form. Even when we understand the reasons for the tissue culture variability we also have to know, from the practical point of view, how to prevent it.

Returning to the question of the headline: Has potato been improved by unconventional techniques? The answer is Yes. It has been proven that the efficiency of potato breeding can be increased by dihaploids (Wenzel et al. 1982) and most German potato breeders use dihaploids today. This has been the first step; the next one will be the incorporation of homozygous monohaploid-derived doubled haploids.

But both fields, the classical and the unconventional one have to be aware of all progress on either side. Only a comparative yield trial can tell whether a new tissue culture-derived resistance can make a new variety. Tissue culture gained in potato already its place in the overall breeding process. Today nobody can estimate the final economic value, but we have to accept that there is an increasing need for food, that the evolution of pests is not slowing down and that agrochemistry reaches limitations; if these parameters accumulate, even an expensive detour will pay in the end.

ACKNOWLEDGEMENT

We thank Dr. Ozias-Akins for critically reading the English manuscript.

REFERENCES

Behnke, M. (1979). Selection of potato callus fro resistance to culture filtrates of *Phytophthora infestans* and regeneration of resistant plants. Theor.Appl.Genet. 55, 69-71

Binding, H., Jain, S.M., Finger, J., Mordhorst, G., Nehls, R., Gressel, J. (1982). Somatic hybridization of an atrazine resistant biotype of *Solanum nigrum* with *Solanum tuberosum* . I. Theor.

136   G. Wenzel et al.

Appl. Genet. 63, 273-277.

Brown, J.S. and Wernsman, E.A. (1982). Nature of reduced productivity
    of anther-derived dihaploid lines of flue-cured tobacco. Crop
    Sci. 22, 1-5.

DePaepe, R., Bleton, D., Gnangbe, F. (1981). Basis and extent of gene-
    tic variability among doubled haploid plants obtained by pol-
    len culture in *Nicotiana sylvestris*. Theor.Appl.Genet. 59,
    177-184.

Foroughi-Wehr, B., Friedt, W., Wenzel, G. (1982). On the genetic im-
    provement of androgenetic haploid formation in *Hordeum vulgare*.
    Theor.Appl.Genet. 62, 233-239.

Hein, T., Przewozny, T., Schieder, O, (1983). Culture and selection of
    somatic hybrids using an auxotrophic cell line. Theor.Appl.
    genet. 64, 119-122.

Hoffmann, F., Thomas, E., Wenzel, G. (1982). Anther culture as a bree-
    ding tool in rape II. Theor.Appl.Genet. 61, 225-232.

Melchers, G., Sacristan, M.D., Holder, A.A. (1978). Somatic hybrid
    plants of potato and tomato regenerated from fused proto-
    plasts. Carlsberg Res.Commun. 43, 2o3-218.

Shepard, J.F. (1982). Regeneration of potato leaf cell protoplasts, in
    Variability in Plants Regenerated from Tissue Culture (E.D.
    Earle and Y.Demarly eds) pp. 47-57, Praeger, New York.

Stabel, P. (1982) Versuche zur cytologisch-morphologischen Chromosomen-
    charakterisierung von Fabaceae, Poaceae und Solanaceae. Dipl-
    Arbeit, Universität Köln.

Stegemann, H., und Löschcke, V. (1976). Index der europäischen Kartof-
    felsorten. Mitt.BBA. 168, 1-215.

Straub, J. (1973). Die genetische Variabilität haploider Petunien. Z.
    Pflanzenzüchtg. 7o, 265-274.

Uhrig, H. (1983). Steigerung der Antherenkultureignung bei einigen di-
    haploiden nematodenresistenten Kartoffelklonen. Vortr. Pflan-
    zenzüchtg. 2, 104-109.

Wenzel, G. and Uhrig, H. (1981). Breeding for nematode and virus-re-
    sistance in potato via anther culture.Theor.Appl.Genet. 59,
    333-34o.

Wenzel, G., Hoffmann,F., Thomas, E. (1977). Increased induction and
    chromosome doubling of androgenetic haploid rye. Theor.Appl.
    genet. 51, 81-86

Wenzel, G., Meyer, C., Przewozny, T., Uhrig, H., Schieder, O. (1982).
    Incorporation of microspore and protoplast techniques into
    potato breeding programs. See Shepard,ibid., pp. 290-302.

DISCUSSION

M.O. KHIDIR: You mentioned that the different ploidy levels between
2n=4x and 2n=2x are stable. Are they stable because they were
asexually propagated? Were they stable when propagated sexually,
particularly those with even chromosome numbers?

G. WENZEL: The even ones are stable propagated vegetatively as well as
sexually. Besides the monoploids, which are not fertile, the
unenploids are in most cases sterile and could be maintained
consequently via tubers.

LONG TERM GOALS IN AGRICULTURAL PLANT IMPROVEMENT

P.R.Day

Plant Breeding Institute,Trumpington, Cambridge, CB2 2LQ, UK.

INTRODUCTION

At the Plant Breeding Institute there are breeding programmes for eight crops; wheat, barley, triticale, sugar beet, potatoes, oilseed rape, field beans (Vicia faba) and red clover. Wheat, barley, oilseed rape and field beans require separate programmes for winter and spring crops. Except for barley, where the two are of equal importance, the winter crop takes precedence and occupies most of the acreage. The crop breeding programmes are designed to compete with the best in Western Europe in producing varieties for UK farmers. They thus serve as a test-bed for new methods and ideas developed by the breeders,and by other scientists who contribute to the technology of breeding. Genetic manipulation lies at the heart of crop plant improvement through breeding and the Institute early recognised the importance of understanding the mechanism of inheritance. In the early fifties it began investigations in cytology and cytogenetics. Six years ago work in cytogenetics was expanded when the ARC implemented a priority programme in plant molecular biology with the objective of exploring and applying the new methods of genetic manipulation to crop plants. We were not alone in this in the UK. The application of genetic manipulation to agriculture is by no means simple. Some problems were caused by a lack of realism in that expectations of early returns were, and in some cases still are, unrealistically high. The first task is to use the science to fashion new tools for the technology of breeding.

In this paper I discuss some of our long term plans for the application of the newer methods of genetic manipulation and molecular genetics to our breeding programmes. First I will briefly review some features of plant breeding that must be taken into account if these goals are to be attained.

INCREASING YIELD

The problem of scale - numbers and time. The major plant breeding goals are to increase yield, improve quality and reduce production costs. Every breeding programme seeks improvements in all three respects. The sustained increases in national average yield in winter wheat since the 50's in the UK have resulted from improvements in agronomy coupled with breeding that has increased harvest index (Austin et al., 1980, Silvey, 1980). An important genetic component was the use of the Norin 10 dwarfing genes Rht-1 and Rht-2 which enabled higher levels of nitrogen fertilizer to be used without loss

DNA sequences concerned. The presence of a particular restriction fragment could then be used as a linked marker for part of a chromosome arm. Both methods are at present too labour intensive to be used except on a small scale. Only a very sophisticated automated system would enable breeders to screen F_2 progenies that may run to tens or hundreds of thousands of plants. Even the analysis of the 2,000 individuals in the F_2 of a single cross by such means could not be justified at present because of the lack of probes or observed polymorphisms that are sufficiently rewarding in terms of their importance to the breeder. This situation could change quite rapidly however as new discoveries are made. Similar logistical problems apply to breeding for improvements in wheat quality by selection of certain high molecular weight glutenins present in the grain endosperm.

Future prospects for yield increase. A central question for wheat breeders is how to continue increasing yield. It is unlikely that harvest index can be further increased from its present level of about 50% without reducing the leaf area and the straw needed to produce and support the grain. During some 80 years of breeding grain yield has more than doubled but there has been no increase in crop biomass (Austin et al., 1980). Further yield increases could thus follow if biomass can be increased. Possible ways of bringing this about include reducing or eliminating photorespiration (Zelitch, 1979), eliminating the alternate, or cyanide insensitive, dark respiration pathway (see Day, 1977) and increasing light saturated rates of photosynthesis (Pmax) (Austin et al., 1982).

The traditional procedure is to seek variation that would contribute to the end in view and try to assemble it in an adapted genetic background. For example the diploid Triticum urartu has a higher Pmax than hexaploid wheat (Austin et al., 1982). Its narrower leaves have more internally exposed mesophyll cell surface area than the larger celled, wider leaved, cultivated wheat thus decreasing the internal resistance to diffusion of CO_2 within the leaf (Parker and Ford, 1982). Selection for narrower leaves with high Pmax coupled with a high leaf area index could lead to an increase in crop growth rates. There is also evidence that chloroplasts extracted from seedling leaves of T. urartu are more efficient in photosynthesis than those of bread wheat (Austin, unpublished).

It is unlikely that the anatomical features of C4 species that promote high rates of Pmax could be introduced into C3 crop plants in the foreseeable future. There is no evidence of any organism with an RuBP carboxylase that does not also have oxygenase activity although there is some variation among organisms in the ratio of carboxylase to oxygenase activities (Somerville, 1983).

When no suitable variation exists in cross compatible related species it can sometimes be produced by treatment with a mutagen. On the whole this has been an unrewarding approach in relation to the large scale efforts made so far. Somatic hybridisation by protoplast fusion will widen the range of forms that can be induced to exchange genetic information. Unfortunately somatic fusion suffers as much, if not more, from hybrid sterility than wide interspecific crosses.

of grain due to crop lodging. However, dwarf wheats suffer in competition with weeds and herbicides are required to control weeds that were kept in check by taller wheats. The use of growth regulating compounds to reduce plant height has similar consequences. As the complexity of the breeder's task increases, the numbers of plants that must be grown for selection in segregating progenies increases in direct proportion. In winter wheat some 700 crosses are made each year and the F_2 populations total some 1.5 million plants. Spring and winter barley programmes are similar in scale. The introduction of a single dominant gene without reducing the selection efficiency for other characters calls for an increase in the size of the F_2 of 33%. A single recessive gene requires a fourfold increase in population size.

Selection in the F_2 and later generations takes account of more than 50 characters that affect the value of the crop to the farmer or the end user. Even so availability of genetic variation has not been a major limiting factor in breeding wheat and barley. For both crops other related species provide sources that may be tapped with varying degrees of ease. In wheat the methods for alien genetic transfer by chromosome engineering developed during the last 30 years have made available reserves of genetic variation (Law, 1983). However, alien sources carry many genes that are unwanted and that must be eliminated by selection.

In wheat, yields are recorded in selections at F_5 and by the time that lines are ready in F_8 for national list testing they will have been in eight yield trials in the previous generation. In the UK, national list trials take two years and recommended list trials two more years. Plant Breeders Rights were introduced in 1964 and impose stringent requirements on the breeder that are designed to protect the farmer. New varieties must not only have demonstrated value in cultivation and use (VCU tests) but must be distinctive, uniform and stable (DUS tests) In practice a new winter wheat variety is not available to the farmer until a minimum of 12 years after the original cross. Production of spring cereals with no vernalisation requirement can be speeded up by carrying out selection and multiplication in the other hemisphere in winter nurseries when three or four years may be saved. Other methods of accelerating the production of uniform lines in the cereals include production of doubled haploids and single seed descent both recently reviewed by Bingham (1983).

Could genetic manipulation help solve these problems? Almost certainly it can in terms of making selection easier by providing methods for detecting the presence of particular genomic DNA sequences or blocks of genes that determine important phenotypic characters such as yield or quality. Increased efficiency in selection could lead to substantial savings in time, labour and growing space. Two methods are available; in situ hybridisation, employing cloned sequences as labelled probes, and restriction fragment length polymorphism. In the latter method variation among segregating plants in the banding patterns of genomic DNA after digestion with several different restriction enzymes would be expected to be associated with phenotypic differences coded by the

It is also non-directed in that much useless genetic information is transferred from the donor species. Some success in obtaining limited gene transfer through irradiation of the pollen used in inter- and intra-specific crosses (Pandey, 1981) suggests that partial gene transfer might also be obtained in protoplast fusion by similar means.

A more exciting long term goal is to recover the genes to be changed by cloning and to attempt to change them in a microbial host. A lot would depend on the design of efficient selection methods. These may be comparatively simple if the plant gene is expressed in the bacterial host but more difficult if it is not. While direct practical application will depend on reimplanting the engineered gene so that it will function in the desired way, the knowledge gained of how to regulate the expression of the plant gene in the comparatively simple background of the micro-organism could be useful in modifying expression in situ in the crop plant. Examples include the regulation of metabolite pathways leading to the formation of valuable secondary metabolites such as seed storage proteins, oils, fats, alkhaloids, steroids etc. In this way the quantity of individual components that contribute to the value of the harvested product might be increased without a corresponding increase in the bulk to be harvested.

Some may question the need to increase cereal and other crop yields since the developed countries presently have a surplus of grain. The US Secretary of Agriculture recently announced legislation to remove 33.3m ha (36%) of US farm land from production in 1983 under the 'payment in kind' (PIK) programme (Block, 1983). Wheat production in the US will be reduced by 20% and corn by 30%. However, the world stock of wheat held mainly in the US, Canada, the EEC and Australia is still less than 20% of the amount consumed each year. The economic magnitude of the surplus and the problems it creates in world trade are real but they seem small in biological terms. The world's population which is now about 4 billion will become 6 billion by the year 2000, less than 17 years away. As the population of developing countries increases so methods of increasing the yields of their own crops must be found. In the developed countries short term economic policies may well reduce the land areas sown to crops now in surplus. Efficient and assured production will be needed on these smaller areas. Research to find new breeding methods will therefore be no less important than before and will serve both developed and less developed nations.

IMPROVING QUALITY

The quality of the crop depends on the presence or absence of certain chemicals in the harvested product. In sugar beet, quality is associated with high extractable sugar content in the root coupled with a low level of juice impurities that complicate refining. Oilseed rape requires oil free of erucic acid and a protein-rich seed meal free of bitter tasting goitrogenic glucosinolates which interfere with feeding to stock. Other crops have comparable requirements. I will illustrate some longterm goals in improving quality in bread wheat.

Grain protein in wheat. Genetic analysis of bread-making quality now depends on two rapid assay procedures: near infrared reflectance analysis (NIR) to measure protein content and the sodium dodecyl sulphate (SDS) sedimentation test for protein quality. Both tests use the same small sample (ca. 10g) of wholemeal flour. Some 15,000 such tests are carried out each year on F_2, F_3 and F_4 progenies (Bingham, 1983).

Although grain protein content in UK wheat may not match that of the best imported North American wheat, much is being done to improve its quality. High molecular weight (HMW) glutenin subunits determine dough elasticity and loaf volume. The molecular weight of the HMW glutenin subunits ranges from 82,000 to 125,000 Daltons. Three complex structural genes Glu-A1, Glu-B1 and Glu-D1 code for the HMW subunits and are located on the long arms of chromosomes 1A, 1B and 1D respectively. Each locus is complex and is composed of several closely linked genes. 'Allelic' differences detected in surveys of 300 wheat varieties revealed 2 patterns of variants for Glu-A1, 11 for Glu-B1 and 6 for Glu-D1 (Law and Payne, 1983). SDS tests of segregants from crosses have demonstrated that the presence of certain HMW subunits is strongly correlated with good bread-making quality. F_2 populations could, in theory, be selected by PAGE analysis of half seeds and sowing only the embryo containing part of those selected. In practice the method is so labour intensive that it is so far only used in the selection of parents but may be useful in later generations, after selection for other traits, when families are smaller.

Another group of endosperm proteins, the gliadins, also contribute to flour quality. These proteins are controlled by six structural genes, but less is known of their role in determining quality than is known for the glutenins.

Extensive collections of wheat relatives have shown a range of HMW glutenin subunits that greatly extend those available in cultivated forms. Chromosome substitution lines with chromosome 1U of Aegilops umbellulata have been prepared to introduce new subunits coded by this chromosome into group 1 chromosomes of wheat.

REDUCING PRODUCTION COSTS

Production costs include the cost of seed, or planting material, soil preparation and cultivation, irrigation and applied chemicals (fertilizers, fungicides, insecticides, nematicides, herbicides, growth regulators, drying agents, and frost protectants). They also arise in harvesting, storing, transporting and processing the crop and in disposing of crop debris when harvesting is finished. All of these costs, and others not listed, can be reduced by the plant breeder. I will review longterm goals for only two, disease resistance and nitrogen fixation.

Disease resistance. Crop protection by breeding for disease resistance is extremely widespread and generally very effective but there are some diseases that cannot yet be satisfactorily controlled because adequate resistance is not available. These include take-all of wheat and barley (<u>Gaeumannomyces graminis</u>); gangrene (<u>Phoma exigua</u> var <u>foveata</u>) and leaf roll virus of potato; and stem canker (<u>Phoma lingam</u>) in oilseed rape (Bingham, 1981). Even when resistance is available and very effective it is often short lived because of the development of new races of the pathogen that are no longer controlled. One strategy which can relieve this problem is to recognise and select for 'durable' resistance which has performed satisfactorily in the field for some years. Another is to deploy crops which have mixtures of genes for resistance. The mixtures may be either multilines or mixtures of varieties with similar maturity dates and end uses (Wolfe, Barrett and Jenkins, 1981). The search for and incorporation of new genes for resistance occupies much of the breeder's time. Genes for disease resistance are thus obvious candidates for gene transfer.

The first goal will be to develop methods for identifying resistance genes present in donor species. These genes appear to operate as sensors detecting the presence of avirulent pathogens and setting in motion resistance mechanisms that result in the suppression of the invading organism. Transposon induced mutagenesis would perhaps be the simplest method to locate genes determining resistance. Other options include genome walking between well characterised markers, the identification of specific mRNAs formed only when resistant tissue is challenged with an avirulent pathogen and the identification of the avirulence gene in a pathogen together with its product to be used as a chemical probe in the host (Day, Barrett and Wolfe, 1983). Since it is likely that there are many different mechanisms for triggering resistant reactions it is unlikely that general methods of wide application will be generated by intensive work on few systems.

A different approach will be to attempt the synthesis by genetic means of resistance mechanisms that are analogues of systemic fungicides. For example the killer polypeptide formed by some dsRNA virus infected strains of the smut <u>Ustilago maydis</u> is toxic to a number of other grass and cereal smuts (Koltin and Day, 1975). It would be worthwhile attempting the synthesis of a cDNA to the dsRNA which codes for the killer polypeptide with the object of introducing this into a host plant to see if it confers resistance to smut. Other candidate genes will be antifungal antibiotics. No doubt the fungicide manufacturers have lists of candidate compounds that have already been tested as potential systemic fungicides but which although non-phytotoxic are either too expensive to manufacture or too ephemeral when applied. Like genes for resistance genes for such compounds will be present and may be active all the time protecting the crop whether or not they are required. Successful methods for introducing new synthetic pathways may also lead to the development of more versatile crop plants where not only the grain will be harvested but the leaves and stems as well if they can be made to synthesise and accumulate valuable new secondary metabolites. A very longterm goal will be a "plant fermentor" analogue where photosynthesis can be coupled directly to the synthesis of important compounds such as hormones, antibiotics and other drugs on an agricultural scale.

Many pathogens are only capable of growing a narrow range of host species. The nature of resistance in non-host plants is largely unexplored and clearly could result from the action of many different genes. Another means of conferring resistance on crop plants may be by transfer of non-host genes. If an efficient method of screening is available, such as a phytotoxin which kills protoplasts or cultured cells, such a shot-gun approach could be worthwhile.

Where resistance is not available and chemical control is uneconomic biological control can be very useful. There is much interest in using the techniques of genetic manipulation for better biological control of insect pests. However, there are prospects for controlling fungal pathogens also, for example by manipulation of dsRNA viruses that determine such characters as hypovirulence.

Nitrogen fixation. Several families of plants have evolved symbiotic associations with prokaryotic micro-organisms which fix atmospheric nitrogen passing it to the plant in the form of ammonia in return for carbohydrate. The best known are the legumes which have root nodules formed by localized invasion by the nitrogen-fixing bacterium Rhizobium. The rates of nitrogen fixation in a legume crop are in excess of 140kg of nitrogen per hectare per year. Farmers now apply on average about 170 kg per hectare of nitrogen to wheat in the UK.

Nitrogen fixation also occurs in a number of unrelated, free-living, soil-inhabiting prokaryotes and in others that are more loosely associated with plant roots than Rhizobium. However, at the present time the best activities under field conditions are poor in comparison with the rates reported for legumes. Unfortunately there are also many denitrifying bacteria in the soil which convert ammonia to N_2 thereby deriving energy and reducing fertility.

Nitrogen fertilizers alone account for 25% of the fossil fuel energy used by arable agriculture and so there is much interest in the prospect of exploiting biological nitrogen fixation (see Hollaender, 1977). Its introduction to non-legumes like wheat, maize and rice by genetic manipulation has been widely discussed. There are many problems still to be solved before this can be achieved but important progress has been made in gaining a better understanding of the genetic organization of the nitrogenase enzyme complex responsible for nitrogen reduction in prokaryotes.

One of the most convenient nitrogen-fixing organisms for genetic studies is Klebsiella pneumoniae found in soil, water and the human intestine. As a genetic tool Klebsiella is almost as convenient to use as E. coli. Since nitrogenase and its genetic controls are very similar in all prokaryotes, there is reason to believe that research findings from Klebsiella will have a wide application.

Recent work summarized by Dixon et al. (1981,1983) shows that the nitrogenase gene cluster occupies a 23 kilobase segment of the Klebsiella chromosome that include 17 genes with identified functions. The genes are organized into eight transcriptional units. Their operation was studied by fusing the E. coli lac operon to the individual promoters so that it responds to the same signals that control the nitrogenase transcriptional units. The lac product, β-galactosidase, is easy to detect. Nitrogenase has two component proteins, one contains molybdenum and iron, the second is an iron-sulphur protein. Both components are irreversibly damaged by oxygen. It is therefore not surprising to find that oxygen represses synthesis of nitrogenase as do ammonia and amino acids. The oxygen sensitivity of nitrogenases is likely to be an important obstacle to the functioning of the gene complex when introduced into a plant cell. Klebsiella only fixes nitrogen in the absence of oxygen. Other prokaryotes, such as Azotobacter, are protected by an active respiration system that can reduce the oxygen levels in cells growing in partially anaerobic conditions. The nodules of legumes contain a special protein pigment, leghaemoglobin, which excludes oxygen from the interior of the nodule. Azotobacter also produces a protein which can protect its nitrogenase from oxygen. However, when the Azotobacter protein is protecting the enzyme the latter is unable to function. It may be possible to recover Klebsiella mutants in which the nitrogenase components are still active but insensitive to the low oxygen tensions likely to be found in root systems.

The gene cluster can be moved from Klebsiella into other enterobacteria where it may or may not function depending on the recipient cells. As soon as a suitable plant transformation vector is available it should be possible to introduce the cluster into plant cells. It is likely that this will only be the beginning of a long period of experimentation to secure function and regulation and to establish the true balance sheet in terms of the high energy costs needed to drive the reaction. These are likely to be equivalent to about 16% of fixed carbohydrate (Austin, personal communication). Hence with a harvest index of 50% the grain yield penalty could be 8%. There would be little point in growing cereals that fix their own nitrogen at the expense of grain production unless the cost of nitrogen fertilizer made it economic.

Since the levels of production needed are likely to be greater than found in legume crops the short term prospects are not good. However, fossil fuel reserves are finite and when they are exhausted biological fixation may well be more attractive if energy costs rise.

Nitrogen fixation by the root nodule bacteria of legumes is very effective but there are indications that it can be made more efficient. In the nodules hydrogen is evolved as a result of nitrogenase activity (see Evans et al. in Hollaender, 1977). Tests of different Rhizobium strains on soybeans and cowpeas indicated that differences in total nitrogen fixed were related to evolution of hydrogen. Rhizobium strains with an active hydrogen uptake system appear to be more efficient since they conserve about one-third of the energy consumed in nitrogen fixation that is otherwise lost through release of H_2. Lim et al. (1980) found that genes for hydrogen uptake (Hup) are by no means widespread in soybean Rhizobium strains so that more stringent selection for Hup should result in the preparation of better commercial inocula. Hup genes in other hydrogen-fixing bacteria are carried on transferable plasmids and might well be transferred into adapted, competitive Rhizobium strains that presently lack them.

CONCLUSIONS

I have tried to be realistic and to restrict my examples to a few that are of interest to my breeder colleagues. There are many others; manipulation of cytoplasmic male sterility is of great interest not only in maize but in sugar beet and oilseed rape. The introduction of herbicide resistance could widen the spectrum of use of important and inexpensive herbicides.

Most of the longterm goals I have discussed depend on the availability of reliable methods for gene transfer. Although these are now becoming available for dicotyledonous hosts of Agrobacterium the cereals still present a major obstacle that must be surmounted. I have referred in discussing disease resistance to the problem of identifying the genes involved. This is a general problem for which methods will be needed that will work in each crop we deal with. We must also learn to cope with the problem of regulating and controlling the genes we introduce. When we can direct where such genes will be integrated in the host genome this may be easier than it seems now. But correct integration also requires a much better knowledge than we have now of chromosome substructure and the mechanisms of development and differentiation.

REFERENCES

Austin, R.B., Bingham, J., Blackwell, R.D., Evans, L.T., Ford, M.A., Morgan, C.L. and Taylor, M. (1980). Genetic improvements in winter wheat yields since 1900 and associated physiological changes. J. Agric. Sci. Cambridge 94, 675-689.

Austin, R.B., Morgan, C.L. Ford, M.A. and Bhagwat, S.C. (1982). Flagleaf photosynthesis of Triticum aestivum and related diploid and tetraploid species. Ann. Bot. 49, 177-189.

Bingham, J. (1981). The achievements of conventional breeding. Phil. Trans. Roy. Soc. London B292, 441-455.

Bingham, J. (1983). Trends in wheat breeding. Proceedings Australian plant breeding conference. Ed. C.J. Driscoll, Waite Agricultural Research Institute, University of Adelaide, South Australia.

Block, J. (1983). Reported in Financial Times p40, Wednesday March 23.

Day, P.R. (1977). Plant Genetics: increasing crop yield. Science 197, 1334-1339.

Day, P.R., Barrett, J.A. and Wolfe, M.S. (1983). The evolution of host-parasite interaction, in Genetic Engineering of Plants. A. Hollaender ed.) Plenum Press, New York, (in press).

Dixon, R.A. (1983). Regulation of transcription of the nitrogen
    fixation operons. Proc. 15th Miami Winter Symposium;
    Advances in Gene Technology: Molecular Genetics of Plants
    and Animals (in press).

Dixon, R.A., Kennedy, K. and Merrick, M. (1981). Genetic control of
    nitrogen fixation, in Genetics as a Tool in Microbiology
    (Glover, S.W. and Hopwood, D.A. eds) pp.   161-185.
    University Press, Cambridge.

Hollaender, A. (ed) 1977.   Genetic   Engineering   for   Nitrogen
    Fixation. 538pp. Plenum Press, New York.

Koltin, Y. and Day, P.R. (1975). Specificity of Ustilago maydis
    killer proteins. Applied Microbiology 30, 694-696.

Law, C.N. (1983). Chromosome engineering in wheat breeding and its
    implication for molecular genetic engineering.   Genetic
    Engineering (Setlow, H. and Hollaender, A., eds) Plenum (in
    press).

Law, C.N. and Payne, P.I. (1983). Genetic aspects of breeding for
    improved grain protein content and type in wheat.   J.
    Cereal Sci. (in press).

Lim, S.T., Anderson, K., Tait, R. and Valentine, R.C. (1980).
    Genetic Engineering in Agriculture: hydrogen uptake (Hup)
    genes. Trends in Biochem. Sci. 5, 167-170.

Pandey, K.K. (1980). Parthenogenic diploidy and egg transformation
    induced by irradiated pollen in Nicotiana. New Zealand
    Journal of Botany 18, 203-207.

Parker, M.L. and Ford, M.A. (1982). The structure of the mesophyll
    of flagleaves in three Triticum species.   Ann. Bot. 49,
    165-176.

Silvey, V. (1981). The contribution of new wheat, barley and oat
    varieties to increasing yield in England and Wales
    1947-78. J. Natl. Inst. Agric. Bot. 399-412.

Somerville, C.R. (1983). Improving photosynthetic CO_2 fixation in
    higher plants. Proc. 15th Miami Winter Symposium; Advances
    in Gene Technology: Molecular Genetics of Plants and
    Animals (in press).

Wolfe, M.S., Barrett, J.A. and Jenkins, J.E.E. (1981). The use of
    cultivar mixtures for disease control in Strategies for the
    Control of Cereal Disease (Jenkyn, J.F. and Plumb R.T. eds)
    pp. 73-80 Blackwell, Oxford.

Zelitch, I. (1979). Photosynthesis and plant productivity. Chem.
    and Eng. News 57, 28-48.

DISCUSSION

E. RUDIGER: Do you believe that you could knock out photorespiration without killing the whole plant? 1) Are mutations conceivable that destroy the oxygenase function of RubCase but leave the carboxylase function intact?   2)   Photorespiration is not just a waste of photosynthetic energy. It apparently has an important 'safety valve' function   by   synchronizing   the   light   and   dark   reactions   of photosynthesis under conditions where the dark reactions are limiting. How about this?

P. DAY: It is possible that mutations that completely eliminate photorespiration will be lethal.   However, since there is some variation among plants in the ratio of carboxylase to oxygenase activity, I think it is worthwhile to try to reduce oxygenase and thereby increase photosynthesis. Of course one way to find out whether photorespiration has a safety valve function would be by demonstrating that it is either impossible to recover oxygenase-less mutants or that they are normally lethal.   An alternative would be to attempt to engineer adjustments elsewhere, perhaps by changes that would make the dark reactions less limiting.

H. JOOS: Do you have any information on the use of Leguminosae in mixed culture with wheat (or other grains) in order to overcome the 'nitrogen costs' in agriculture? This biological way of solving this problem has at least several other advantages like improvement of soil structure, less use of herbicides (the weeds that might develop are in competition with the Leguminosae), improvement of the resulting straw which can be used as animal food, etc.

P. DAY: My colleague Martin Wolfe has experimented with mixtures of peas (Pisum sativum) with wheat and barley. Small-scale experiments with field beans and barley suggest that mixed cultures can be more productive than monocultures (see Martin and Snaydon, Expl. Agric. 18, 189, 1982; and J. Appl. Ecol. 19, 263, 1982).   They also have advantages in the control of diseases. Wheat pathogens that land on the legume are unable to infect it and vice versa.

P. STARLINGER: Could you comment on the desirability of increasing phosphorus uptake, perhaps by manipulating mycorrhizae.   Will phosphorus become a scarce resource in the foreseeable future?

P. DAY: The role of vesicular-arbuscular mycorrhizae in increasing the effective absorption area of root systems has been neglected by breeders. There are some very striking examples of the effectiveness of introducing mycorrhizal inoculum in establishing good plant growth. Some years ago there was much concern over phosphate fertilizer reserves. Alternative sources and new deposits have rendered this concern less pressing, but it could become an important limiting factor in the future.

P. STARLINGER: Could you comment on the desirability of having nitrogen-fixing cereals? Given the energy consumption of this process and its probable negative effect on yield, it might be more useful to

use nitrogen fertilizer to maximise yield than to sacrifice yield in
order to perform bioconversion of solar energy on fertile soil.   Is
this reasoning faulty and, if so, in which respect?

P. DAY: If the cost of having nitrogen-fixing cereals is a gross
reduction in yield, this cost would only be met when the energy cost of
producing nitrogen fertilizer makes its application to our fields
increasingly uneconomic. In the meantime, of course, if we are able to
increase photosynthetic yield to offset this additional cost, we may
be able to re-write the equation.

VAN DEN DAELE: The application of genetic manipulation to human needs
poses the problem of how the priorities for such applications are
determined. A highly ambivalent example seems to be the prospect that
herbicide resistance is transferred to crop plants.   This might
further enhance capital intensive use of chemicals in agriculture.
Given the fact that the chemical load on our environment already poses
ecological problems, is this not a step in the wrong direction —
reflecting more the commercial interests of the chemical industries
rather than the needs of agriculture?

P. DAY: The genetic manipulation of herbicide resistance could have
considerable   advantages   in   making   available   effective,   cheap
herbicides that cannot be used on certain crops because of their
sensitivity. Over the long term, I would agree that to increase our
dependence on applied chemicals for agricultural production is
undesirable, not only because of effects on the environment but
because of the additional costs.   There has been some interest in
exploiting the properties of allelopathy, namely the synthesis by
certain species of compounds which inhibit the growth of other plants
competing for the same ecological niche. It may be possible to create
new forms of crop plants which secrete allelophatic compounds that do
not make them toxic to human and other animals, thus reducing our
present dependence on herbicides.

GLOBAL ACTIVITIES IN PLANT GENETIC RESOURCES CONSERVATION

J.T. Williams

International Board for Plant Genetic Resources, FAO, Rome, Italy.

INTRODUCTION

An international programme to collect and conserve the genetic resources of crop plants was only established a little under 10 years ago. This had been considered a task for high priority action for at least a decade beforehand, largely due to the efforts of FAO and its Panel of Experts chaired by Sir Otto Frankel. The International Board for Plant Genetic Resources (IBPGR) was created in 1974 and has been actively involved since then in establishing and developing a global network of genetic resources activities.

The IBPGR is a centre of the Consultative Group on International Agricultural Research (CGIAR), an informal group of donors co-sponsored by FAO, UNDP and the World Bank. The CGIAR mobilizes funding for the IBPGR. The IBPGR aims principally to act as a catalyst to stimulate action, to coordinate the activities world-wide and to fund such activities which are considered urgent.

To date, research centres and scientists in almost 100 countries are collaborating with the Board. It has designated a world network of institutions to act as base storage centres (i.e. for long-term maintenance) for stocks of the major seed crops and currently this network consists of 38 centres in 29 countries for 30 crops or groups of crops that produce seeds which can be dried and held at low temperatures. It has just commenced a similar exercise for the designation of centres to hold clonally propagated crops and will, in the near future, attempt to identify and formalize the many links between base centres and those where scientists are active in utilizing materials and carrying out other tasks such as the screening of collections for special purposes, the routine characterization of materials and the documentation of relevant data.

PRACTICAL CONSIDERATIONS

This short resumé (amplified in numerous reports, e.g. CGIAR, 1982, IBPGR 1983a, b,) sounds very simple but there were many practical

problems to overcome at the outset and a number of these have not
yet been solved.

To start with there was inadequate information available 10 years
ago on the diversity of even the major crop species, other than
direct extrapolation of the concepts of Vavilov laid down many years
previously and a mixture of generalizations based on imperfectly
understood ecologies and somewhat better known taxonomies. With
limited funds it was necessary to rank the crops in a priority order
for action and to seek information on known patterns of variability.
The task was urgent: landraces of crops were being swept away at
a rapid rate due to the advances in agriculture, particularly in the
developing world, and concern was being expressed about the very
survival of many related wild species.

Through the collation of detailed advice of crop scientists the
world over, the Board now has to hand the best available information
for most of the important crop plants. A recent planning committee
meeting of the IBPGR suggested that the collecting programme (see
below) should be extended to keep scientists in the field for longer
periods and for them to conduct sound eco-geographical surveys.
This would certainly expand our information base.

Secondly the practicalities of field collecting in vast regions of
the world posed many organizational problems. Just over 10 years
ago we all thought that institutions and scientists would sooner or
later accept the challenges and carry out the work. In practice the
Board's Secretariat had to slowly change its structure and become
not only the Board's scientific and administrative arm but also its
major operational arm. Through these activities the Board has, to
date, arranged and helped to finance some 250 collecting missions
in more than 70 countries. There is still much to be done,
especially in remote areas. To cope with this work the Secretariat
had to expand and develop its own operational bases in important
regions of the world. The regions are not clearly demarcated like
the original Vavilovian centres of origin but cover most areas of
the world apart from the frozen wastes. The need for action in many
parts of the world has become more apparent as the Board's
responsibilities have been extended to include, in addition to the
staple food crops, vegetables, woody species for fuelwood and
environmental stabilization and forages (IBPGR, 1981).

The IBPGR has interpreted its mandate in a conservative way by
placing emphasis on major food crop species. It is not concerned
with forestry. One major reason for this is because it is a centre
of the CGIAR. Nonetheless, it does include in its priorities over

50 species for first priority action either globally or in some specific region.

Operationally it tries to encourage national programmes in key areas and to forge cooperative linkages within regions. Ultimately however, the Board is concerned with crops and its approaches remain pragmatic. In some parts of the world a regional approach can be taken, in others it would not be successful.

The funding for the work of the IBPGR is allocated in a logical way among crops. About half the funds relate to cereals and food legumes; there are other allocations for other groups of crops. The IBPGR will probably not be able to work in the depth needed on minor crops and such groupings as medicinal plants. The national efforts will need to take care of the species just as in the case of those for which there should be in situ conservation. The Board has not relegated all responsibility in this area; it commissioned a report on in situ conservation which started a new look at the problems and it is ready to provide advice on, for instance, inventory of wild species, but will not have the funding to support the work.

The IBPGR operates on core funding of a little under US$4 million. Multilateral agencies spent about $2.78 million in 1982 and bilateral aid programmes about $3 million on genetic resources. The CGIAR provided additional funds for the germplasm work of the International Centres at a level just over $9 million. However, in total this is not a large sum for global activities.

Thirdly when we started operations there was only a handful of gene-banks in existence. One of the most important developments has been the increase in these facilities, particularly in the developing countries. Fig. 1 shows the numbers suitable for long-term storage. There are other centres with medium-term storage which are involved with evaluation, documentation and exchange.

OTHER ASPECTS OF THE PROGRAMME

The Board does not concern itself only with collection and conservation. The programme includes support to:

(i) fund research on seed physiology and other conservation techniques so that the best practical procedures for curators can be determined;

(ii) provide an information service on existing collections of germplasm;

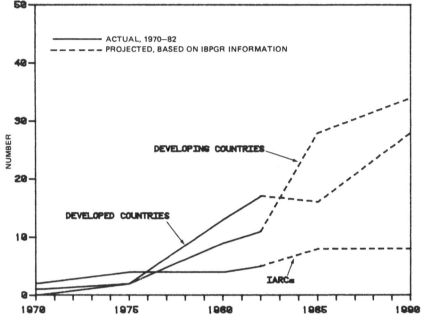

Figure 1.   The number of genebanks for long-term storage of crop germplasm
that were in operation between 1970 and 1982 and projected to 1990.

(iii)   stimulate and support the characterization and evaluation
of material using standardized descriptors which have been
developed in order to enable information about samples to
be catalogued in a standard form;

(iv)   fill a big gap in the availability of trained manpower.

The Board's evolving role is expected in the immediate future to move
strongly into support of research and development of _in vitro_
techniques for genetic conservation and in the mid-term to the
documentation of samples according to gene symbols.

We wish to extend our dialogue with scientists involved with genetic
manipulation to see how our mutual interests can be furthered.   In
this context the following will be of interest:

1.   In relation to _in vitro_ techniques there are no methods of cryopreservation available for any crop species which can ensure almost
total recovery, although there are promising signs (Withers, 1980;
IBPGR, 1983c).   Of grave import is the fact that genetic stability
has been addressed only in a cursory way.   We are particularly
interested in maintaining existing variability and not to release
new variablity, although this is of current interest to breeders

(Larkin and Scowcroft, 1981). Associated with long-term maintenance of vegetatively propagated species because of the international aspects of our work and the need to move stocks across quarantine barriers we shall have to be involved with disease indexing. Monoclonal antibody technology can be used to produce antibodies for diagnostic purposes. At present the agricultural community needs diagnostic antisera to identify viral, bacterial and fungal pathogens particularly for rice, maize, potato, cassava, citrus and fruit trees but we are a long way from developing countries using the techniques. The IBPGR has recently identified high international priority research needs on disease indexing of cassava, sweet potato, yam, aroids, banana, sugarcane and temperate fruits (IBPGR, 1983c).

Moreover, the techniques developed recently for biochemical markers will assume great importance in our work. We have a lot to learn in this area from other scientists. We must be willing to speak similar languages across the borders of the biological subject areas. To illustrate how time could be wasted without this, I have only to elaborate on the stated need from the 1960s to date for research on in vitro techniques for the long-term conservation and genetic stability of stocks of clonal materials. There has been almost no relevant work in this area although research was helped tremendously for the cleaning up of stocks through culturing and the rapid propagation of materials and their easier exchange. However, even here, biochemical work is needed to ensure that supposedly clean cultures still do not contain viroids.

2. On the information side we have found it extremely difficult to convince curators and scientists of the urgent need to document data. In the first place, data are just not available for the majority of the samples in existing collections. We do what we can but a massive emphasis is needed to make the materials more amenable for use. We have a full-time officer trying to sort out the collections of wheat and that alone is a tremendous task. Perhaps a recent surge of interest in pre-breeding will help our educational task: but we have to be aware of the constraints. To start with, data aspects were and still largely are relegated to a low priority compared to the so-called urgency of collecting. But, so too have the careful maintenance of the materials and multiplication and regeneration according to good scientific standards (see Fig. 2). This is probably due to the fact that most of the genebanks of the world have only been built in the past five years. Most institutes and indeed nations, underestimate the costs and the staffing necessary for the scientific genetic resources work. This may sound somewhat harsh, but having been involved with the administration of funding for national genebanks in 20 developing countries, I can

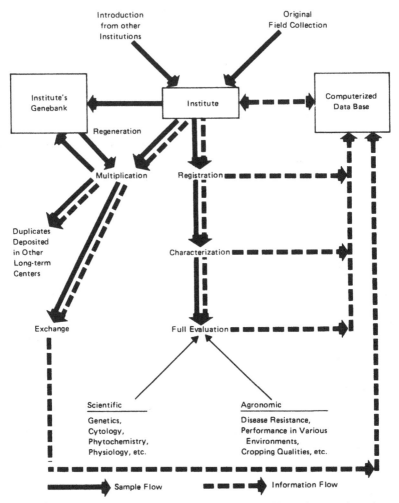

Figure 2. Procedures in evaluation and handling of accessions in crop germplasm banks.

speak with authority.

3. Despite these constraints, never before have such large collections been available to breeders. They will be built up to make them more comprehensive, particularly in their representation of wild species. There are clear possibilities in breeding using wide crosses; and genetic manipulation should make striking advances here. Breeders have been very traditional in many ways, frequently seeking specific genes from the collections but advances can be made which might result in spectacular short-term gains. Wild relatives of crop plants are under-exploited because breeders

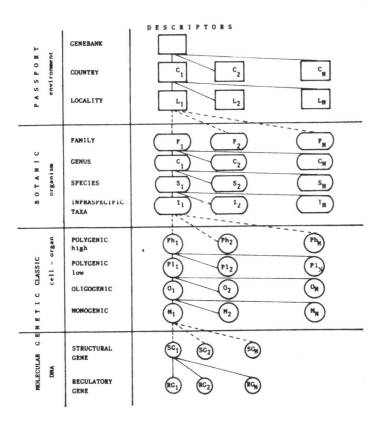

Figure 3. A hierarchical scheme for descriptors

genotypes evolved. The programme I have discussed is concerned with
crop species. There are many wild plants that could contribute to
man's welfare and health. Plant breeders and biotechnologists have
collections of genotypes available for improving or modifying crop
plants. Few have collections to hand of medicinal plants.
Similarly, forest species have not received the attention they
should.

I started off by saying that an international programme on crop
germplasm conservation was fraught with scientific and operational
problems. Most of these can be solved. However, many issues are
now becoming political. FAO and the IBPGR - as well as other
International Centres of the CGIAR with crop germplasm programmes
- have worked on the principle of free availability of materials.

are frequently unfamiliar with them;   they prefer to intercross
elite materials and do not wish to have to deal with sterilities and
deleterious genes. However, wild species have co-evolved with pests
and diseases and are a valuable part of the genepool. Apart from
their utilization by traditional means we now have available in
vitro methods of pollination and fertilization, hybrid embryo culture
and fertilized ovule or ovary culture as well as fusion of somatic
cells and protoplasts from two species (Keller et al., 1982). There
is room for optimism for the future (Collins, 1982).

4. In relation to documentation, genetic conservationists have been
involved with gross characteristics and genotype/environment interaction in their expression. The genetic basis of resistance to
environmental stresses such as heat, drought, salt soils and others
are imperfectly understood. A greater understanding could be gained
by more intensive work on in vitro cultures. There are many genes
determining plant characteristics and much more information is
required about the specifics of gene organization and the
developmental regulation of gene coding for specific traits. Work
on soyabean seed proteins at UCLA, USA, has shown the value of this
type of research.

In a model developed recently in conjunction with Stig Blixt (1982),
particularly for a cooperative programme of more than 20 European
countries, we have considered a hierarchical descriptor scheme
extending to the genic level. At the higher levels are taxa and
various genetically based descriptors whether polygenic, oligogenic
or monogenic (Fig. 3). Descriptors have to be logical and based on
sound science. For instance, highly polygenic seed colour states
of peas could be divided into thousands of descriptor states and
such continuous truly polygenic variation has to be coded sensibly.
If it has been difficult to convince scientists of the need for
lists of minimal descriptors and the necessity for basic electronic
documentation systems it will be even more difficult to ask them to
expand their horizons to take account of genes. But it will come,
although it will take time just as the direct application of
molecular genetics to the development of new cultivars of·crop
plants will take time.

        POSTSCRIPT

There are many reasons why top priority should be given to
protecting the genetic resources we have now. The message has been
spread widely but not widely enough at the policy-maker level to
have the right magnitude of support. Efforts in the tropics are
particularly inadequate where major efforts for agricultural
increase are imperative and where many of the crop species and

The member countries of FAO are now asking for this to be formalized and that some type of legal framework acts as an umbrella for genetic resources activities. We hope that this does not inadvertently provide further constraints.

REFERENCES

Blixt, S. (1982). The pea model for documentation of genetic resources, in Documentation of Genetic Resources: A Model, IBPGR.

CGIAR (1982). 1982 Report on the Consultative Group and the International Agricultural Research it supports: an Integrative Report. ICW/82/06, CGIAR, Washington, D.C.

Collins, G.B. (1982). Plant cell and tissue culture: an overview, in Priorities in Biotechnology Research for International Development, pp. 230-253, National Academy Press, Washington, D.C.

IBPGR (1981). Revised Priorities among Crops and Regions. AGP:IBPGR/81/34, Rome.

IBPGR (1983a). Annual Report for 1982. (In press), Rome.

IBPGR (1983b). Facts about the IBPGR. Rome.

IBPGR (1983c). Report of the First Meeting of the In Vitro Committee. (In press), Rome.

Keller, W.A. et. al., (1982). Production, characterization and utilization of somatic hybrids of higher plants, in Application of Plant Cell and Tissue Culture to Agriculture and Industry (D.T. Tomes et al., eds.), pp. 81-114, University of Guelph Press, Guelph.

Larkin, P.J. and Scowcroft, W.R. (1981). Somaclonal variation - a novel source of variability from cell cultures for plant improvement. Theoret. & Appl. Genetics, 60, 197-214.

Withers, L.A. (1980). Tissue Culture Storage for Genetic Conservation. AGP:IBPGR/80/8, Rome.

DISCUSSION

G. WENZEL: Would it not be more feasible to collect genes especially in vegetative propagated crops and out-breeders, e.g. in the form of apple seeds or other heterozygous sources, rather than to collect specific genotypes?

T. WILLIAMS: It is much more feasible in clonal materials but the breeders at the present want collections of clones readily available. Both methods should be used.

O. FRANKEL: Dr. Williams referred to the FAO Expert Group on genetic resources acting effectively as a pressure group. While this is true to a degree, it was inevitable because the conservationists, who now so loudly proclaim their concern for genetic resources, were entirely silent when their help would have been valuable. But the FAO group did more. It laid the foundation for our scientific knowledge of genetic resources, and wrote the two books on the subject which became the basis for genetic resources work.

GENETIC PERSPECTIVES OF GERMPLASM CONSERVATION

O.H. Frankel

CSIRO Division of Plant Industry, P.O. Box 1600, Canberra City, 2601, Australia.

ABSTRACT

Curators of large germplasm collections evaluate them mainly for "observable" characters. Breeders of intensively bred crops draw on these collections only when their "working collections" of adapted types fail to fill specific requirements. In developing countries indigenous varieties and their present day derivatives remain the principal genetic resources. The scale of evaluation needs to be reduced if in-depth research and evaluation, and through it utilization, are to be encouraged. This could be achieved by a rationalization of evaluation through the use of biogeographical information, combined with a drastic reduction of redundancy. The proposed pruning would result in a "core collection" for distribution and evaluation, the remainder being a "reserve collection" for research and for future needs. There is also a case for reviewing the over-elaborate and over-centralized evaluation and information systems which have been introduced. Emphasis should be shifted to genetic analysis as a basis of gene transfer, especially by molecular techniques. In turn, the latter may be used to isolate genetic components of yield. Much more attention is likely to shift from land races to wild relatives, not only as sources of resistance to restraints, but for a much needed expansion of the gene pool.

1. GENETIC RESOURCES - THE GENETIC PHASE.

Since the beginning of the genetic resources era some sixty years ago, three phases can be discerned:

1st phase - with the impulse from Vavilov's discovery of centres of genetic diversity, emphasis on biogeography, taxonomy, evolution.

2nd phase - with the impulse from "genetic erosion", emphasis on conservation (especially of land races).

3rd phase - with the impulse from utilization, emphasis on genetics.

In this third phase the emphasis is shifting from salvage and preservation of diversity in the field, to study and use of genetic diversity now assembled in germplasm collections. This paper addresses the question to what extent genetic thinking, genetic evidence, and genetic methodology can assist in:

(i)   identification and characterization of genetic components of value in current plant breeding,

(ii)   developing principles for safeguarding resources for the future,

(iii)   developing guidelines for the organization and structuring of germplasm collections that will raise the effectiveness of 1 and 2.

2.   GENETIC BUILDING BLOCKS - "OBSERVABLE" AND "COMPLEX" CHARACTERS.

Germplasm collections should, and to a degree do, reflect the requirements of plant breeders of the day, notwithstanding the responsibility for preserving resources for future needs which cannot be foreseen. Plant breeders use two kinds of genetic components which are functionally, and to a large$_*$ degree genetically, distinct, though both have impacts on productivity .

(i)   Characteristics which are phenotypically readily identifiable, though this may require specific environmental conditions. Many years ago I called them "observable characters" (Frankel, 1947). They are either based on a small number of genes or, if polygenic, they are strongly inherited. Examples are resistance to diseases and pests, tolerance of adverse climatic or edaphic factors (see Table 1), and various morphological and physiological characters related to productivity.

(ii)   Complex (or quantitative) characters, which are subject to environmental variation and are polygenic. They are largely responsible for adaptation and productivity in specific environments. Most plant breeders will agree with Simmonds (1983) that "most economically important traits are polygenic".

Though not formally acknowledged, this distinction in function and genetic constitution has been recognized by curators and by plant breeders. Curators of the large collections which constitute the network of genetic resources centres co-ordinated by the International Board for Plant Genetic Resources (IBPGR) evaluate their collection mainly for characters of category 1. An example is the evaluation of the germplasm collection of rice of the International Rice Research Institute (IRRI), probably the most representative collection of any major crop. Through its own plant breeding programme and its collaboration throughout Asia, IRRI is in close touch with rice breeders and their requirements. It is therefore significant that the IRRI Genetic Evaluation and Utilization Programme (GEU) concentrates on characteristics of category 1, i.e. characters which are "observable" under suitable conditions (Table 1) (Khush and Coffman, 1977). This programme continues to supply IRRI's and other rice breeders with a stream of valuable germplasm.

The use that plant breeders make of germplasm collections is not nearly as well documented. There are statistics of numbers of

---

*   Throughout this paper, "productivity" is used in the widest sense including both biological and economic aspects.

samples supplied, but not of their use and effectiveness. Hence a survey of breeding materials used by a representative group of US breeders of five major crops is particularly helpful (Duvick, 1981). It shows that the majority of breeders prevailingly use adapted, high-performing varieties or breeding lines - their own or other breeders' - as sources of genetic diversity and, surprisingly, even as sources of resistance to pathogens and of tolerance of environmental stress. When these fail they resort to germplasm collections. Thus, subject to the now recognized need for diversification (see the next paragraph but one) breeders of these highly developed crops tend to confine themselves to their "working collections" which consist largely of adapted material. After all, it is through the use of resources of this kind that breeders have been able to maintain, and in some crops steadily to increase the yield potential over many years, and continue to do so (Evans, 1983).

The position is different in less developed countries, where gene pools have not yet been narrowed by past selection and breeders look to germplasm collections as sources of yield improvement as well as of resistance to yield restraints. Here national collections, largely consisting of locally adapted land races and their derivatives, may take the place of breeders' working collections.

We have stated that both curators of large collections and breeders of extensively bred crops evaluate germplasm collections mainly for observable characters. While this is true it also is an oversimplification, since it overlooks two important aspects in the role of germplasm collections. The first is their evolutionary role as reservoirs of genetic diversity. The dangers of genetic homogeneity are increasingly realized, as is the potential for yield improvement through the inclusion of "exotic" germplasm in breeders' gene pools. This idea has led to programmes such as the co-operative project of CIMMYT (International Centre for Maize and Wheat) and Oregon State University to transfer useful genes between spring and winter wheats, and is applied in other programmes of CIMMYT. The use of diverse germplasm sources is by no means a new departure. It was strongly advocated by Vavilov half a century ago (personal communication, 1935).

The second aspect is that evaluation is by no means confined to the large collections. Mainly, but not exclusively in developing countries, locally based collections with direct contacts with plant breeders, and indeed the breeders themselves play a vital part in evaluation. This is further discussed in section 4.

3.   EVALUATION FOR USE.

Three closely interacting steps can be recognized:

(i)   Identification of characters relevant to current plant breeding.

(ii) Genetic characterization of such characters. Genetic information facilitates conventional transfer and is a precondition for chromosome manipulation and for genetic engineering involving DNA transfer.

(iii) Reduction in the number of accessions by minimizing redundancy, to encourage research and evaluation in depth as a basis for extensive utilization. The main emphasis in this discussion is on the third step.

Germplasm collections contain three categories of germplasm which are ecologically, historically and genetically distinct:

(a) land races - old varieties used prior to the availability of modern varieties. Characteristics: adaptation to local environment; genetic diversity between and within land race populations.

(b) wild relatives of crop species.

(c) modern varieties (current and obsolete).

In the last fifteen years the main emphasis has been on land races because of the realization that their existence was seriously threatened by the spread of modern varieties. Many of the largest collections, such as that of IRRI, consist prevailingly of land race material, as do collections of many less developed crops, especially in the tropics. As Table 1 shows, the IRRI collection has yielded a wealth of valuable genes which have transformed the prospects of rice breeding and cultivation. So may collections of other crops once they are as intensively evaluated as IRRI's, provided they are as widely representative.

Table 1.   Land races of rice collected in South and South East Asia, with special traits identified in evaluation programmes (from Chang, 1980).

Reported features	Samples (no.)		(no.)
Tolerance for salinity	343	Resistance to nematodes	5
Tolerance for aluminum toxicity	290	Disease resistance	1634
Tolerance for acid soils (wetland)	216	Insect resistance	619
Tolerance for alkaline soils	324	Aromatic types	120
Tolerance for iron toxicity	16	Multiple tolerances	62
Tolerance for iron deficiency	3	Semidwarfs	84
Tolerance for phosphorus deficiency	3	Nonpreference by rodents	10
Dryland and drought-resistant types	2781	Medicinal purposes	13
Floating and flood-tolerant types	776		
High-elevation and cool-tolerant types	671	Total	7970

Evaluation, perhaps more than any other genetic resources operation, calls for an "inverse economy of scale", not only to save funds, effort and time but because smaller numbers make possible an intensification of evaluation and research. It is feasible to search

the 60,000 rices held at IRRI, or the 70,000 wheats of the Vavilov
Institute in Leningrad, for a simple morphological or biochemical
character, but a meaningful evaluation of physiological attributes
calls for a greatly narrowed scale, based on some form of
discrimination. I shall discuss two approaches. The first seeks to
narrow the scope of evaluation on the basis of biogeographical and
ecological information, the second on the basis of genetic redundancy
of accessions.

Biogeographical and bioclimatic information is as basic in evaluation
as it is in collecting. For tolerance of waterlogging or of drought
one turns to environments exposed to these conditions. Similarly,
the distribution of a pathogen provides a lead to likely sources of
resistance. Dinoor (1970) found that resistance to Puccinia coronata
of Avena sterilis was considerably more frequent in the mesic north
of Israel than in the arid south. Frequencies of resistance of Avena
fatua to P. coronata in New South Wales has a similar trend,
declining from north to south, in line with the temperature and
humidity gradient, though this is less marked for P. graminis
(Burdon et al., 1983). Oates et al. (1983) found a cline for racial
diversity and for virulence in the two rust species which paralleled
the cline for resistance found in A. fatua.

Observations of this kind are few and far between. In their absence
it seems reasonable in the first instance to restrict evaluation to
accessions from areas where the yield restraining factor - be it a
pathogen or a soil component - is likely to be prevalent.

Some information can be gleaned from the collections themselves,
although most of the older accessions are poorly documented for
biological status and place of origin (Frankel and Soulé, 1981, pp.
185-189). The most useful information on genetic variation would
come from studies of inter- and intra-population variances in land
races and wild relatives. Very little is known about land races, but
more about wild relatives. Allozyme surveys of wild species of
Lycopersicon (Rick et al., 1977) and Hordeum (Brown et al., 1978)
have revealed extensive clinal, inter- and intra-population
differentiation. This supports the sampling procedures proposed by
Marshall and Brown (1975) which emphasize the number and diversity
of sampling sites.

Yet all too little is known about the distribution of characteristics -
and alleles - of direct value in plant breeding. What is the
distribution of resistance genes such as Sr 26 from Agropyron
elongatum which confers resistance to stem rust (P. graminis tritici)
in wheat? Is it widespread in some regions, or sporadic? That such
alleles are found among a few samples of a species points to the
former. The subspecies yanninicum of subterranean clover
(Trifolium subterraneum L.) grows in waterlogged conditions in the
Balkans, and one of its ecotypes is a useful pasture legume in
Australia (Katznelson and Morley, 1965). Unfortunately it has a
high content of the oestrogenic isoflavone formononetin which induces
infertility in ewes. In a collection of T. subterraneum made by
Katznelson in Greece and Yugoslavia, there were 13 samples of the

rare ssp. yanninicum, and among the 38 plants extracted from them, at least one had the sought for combination of two complex characters, low oestrogen content and early flowering (Morley, personal communication).   As a combination of two characters this suggests fairly high frequencies in the small populations involved. We need to know much more about the population structure of wild relatives which are becoming increasingly prominent in utilization; for land races of some of the major crops it is almost too late. There is an urgent need for genecological and population genetic studies of characters of the kind used by plant breeders as a guide to biologically informed sampling.

This leads to the second approach towards a rationalization of evaluation, which seeks to reduce redundancy in collections.   Two aspects are involved.   The first is redundancy arising from the identity or close similarity of two or more accessions. This has been raised by Frankel and Soulé (1981, pp. 243-244 and 249-251) and further elaborated by Marshall and Brown (1981). The second aspect is an apparent overrepresentation in sections of the collection.   Criteria for defining desirable or excessive levels would include ecological information, collection or "passport" data, and results of preliminary evaluation, gradually reinforced by genecological evidence.   If overrepresentation is ascertained stochastic reduction would be applied.

Clearly, such thoughts are contrary to the dogma that the larger the number, the greater the chance of finding what one is looking for. This may be justified for rare and for sporadically occurring alleles, but not for the locally advantageous ones which are our preferential target (Marshall and Brown, 1975).   My hunch is that to find such alleles will not require the dense sampling pattern now recommended, but that they will be found in a small number of sites within an ecologically defined target area.   This is likely to be the case in land races and, as previously suggested, it may be common in wild species.

Finally, whatever the nature of redundancy, the loss of genetic variation is minimized if the accessions involved, rather than being differentially retained, are bulked to form a subsample (Frankel and Soulé, 1981; Marshall and Brown, 1981).   Some of the matters raised in the last two paragraphs will be more fully discussed elsewhere.

4. A NEW LOOK AT SIZE, ORGANIZATION AND EVALUATION.

So far we have presented evaluation as being restricted to the large collections. In fact, what is perhaps the most significant evaluation is done in national or specialized collections and by the breeders themselves.   It is only in the local environment that the potential of adapted and exotic stocks and their interaction can be unequivocally assessed.   Thus evaluation is a multiple process, carried out not in one,   but   in   a   number   of   sites   over   periods   of   years. The need for economy of size is thus multiplied.   It is evident that it is not size as such but genuine diversity which needs to be emphasized if collections are to be widely used.   Procedures like those suggested in the preceding section would need to be applied determinedly if this end is to be achieved.   This pruning procedure would result in assembling accessions representing, with a minimum of repetitiveness, the genetic diversity of a crop species and its relatives. For this assembly I propose the term "core collection", the term "active collection" having been pre-empted.

It would however, be imprudent to jettison, without cogent reason, accessions not included in core collections, except duplicates that are convincingly identified as such. The full range of accessions should be retained:

(i)   for population genetic or other kinds of research,

(ii)   as a source for low-frequency alleles,

(iii) as a reserve for unforeseeable future needs.

These accessions would form the "reserve collection", which, together with the core collection, would constitute the germplasm collection. The core collection would be the main source for (i) distribution to plant breeders and to national collections, and (ii) for evaluation. The reserve collection would be available for specialized or research needs. Material could be shifted from one to the other in response to increasing information.

National collections could with advantage adopt a similar policy, subject to their responsibility to safeguard a fuller representation of local land races and crop related wild species than would be possible for an international collection. Most national genetic resources centres are now equipped with cold-storage facilities. Breeders' working collections would be strengthened by the influx of accessions resulting from intensified evaluation. Perhaps of equal importance, a streamlining of collections through concentration on ecological and genetic distinctiveness will help in the selection of materials for physiological and genetic studies of components of adaptation.

Finally, there is a need for a new look at evaluation and documentation. I freely admit, as quoted by Marshall and Brown (1981), to having advocated systematic evaluation and documentation as a precondition of utilization. This has been widely accepted and may have contributed to the development of centralized information systems with large numbers of descriptors and descriptor states, compiled by specialist groups under the auspices of IBPGR. In addition to necessary information on the origin of an accession they contain a detailed morphological description - in rice 43 descriptors, with up to 10 descriptor states (IBPGR-IRRI, 1980) - followed by data on development, reaction to biotic factors, and grain characteristics. Marshall and Brown (1981) question what benefit breeders would derive from the elaborate morphological characterization, and suggest it could be useful for the identification of duplications, but hardly for any other purpose. I tend to concur. The authors further contend that evalution of real value to breeders needs to be done in their environment and preferably by themselves. Even data on disease resistance may have no more than indicative value unless pathotypes at test and breeding sites coincide. Is there redundancy also in evaluation and documentation of germplasm collections?

5.   PERSPECTIVES

(i)   Wild relatives.   The   effectiveness   and   versatility   of   wild
relatives as genetic resources of crop species has been emphasized
by Harlan (1976) and others.   They are certain to be used much
more   widely,   and   not   only   as   donors   of   specific   alleles   or
characteristics, of which the tomato is an outstanding example (Rick,
1982).   Unexpectedly,   marked   increases   in   yield   were   found   in
crosses of domesticated oats (Avena sativa) and its wild progenitor,
A. sterilis (Frey, 1981), and there are similar indications in other
species crosses (Harlan, 1976).   As Frey points out, the widening of
the gene pool has raised the yield potential of oats above the plateau
which had been reached, with the prospect of further increases.
This   is   an   added   incentive   for   assembling   and   studying   a
representation   of   crop   relatives.   Of   even   more   far-reaching
significance for the use of alien germplasm could be the application
of molecular techniques for gene transfer which are rapidly evolving
(see chapter by Peacock in this book).   Two aspects are of prime
importance   here.   First,   transfer   can   be   effected   with   great
precision,   without   the   burden   of   useless   or   harmful   background
which has proved a major hurdle in all other forms of gene transfer.
Secondly,   transfer   can   occur   beyond   the   boundaries   of   sexual
compatibility which suggests the possibility of extending gene pools
to distantly related taxa, and beyond.

(ii)   Somaclonal variation is discussed in the chapter by Green in
this book.   It is likely to make a valuable contribution to the
breeder's   genetic   resources,   in   the   first   instance   perhaps   in
resistance breeding.   In analogy with mutation breeding, somaclonal
variation conserves the original plant type, whereas in conventional
breeding   its   restoration   requires   generations   of   selection   or
backcrossing.   But the implications of cell genetics for the future of
genetic resources may go far beyond such modest beginnings.

(iii)   Genetic analysis.   The need for genetic information at the level
of the gene, the individual, the population, and the ecological
community has been stressed throughout this paper.   A great deal of
information   on   the   genetics   of   "observable"   characters   exists,
extending   from   innumerable   resistances   to   pathogens   and   their
biotypes,   to   physiological   characteristics   such   as   vernalization
requirement, photoperiod response, or metal ion toxicity.   But the
most important question - the genetic components of yield - remains
unanswered,   being   too   complex   for   formal   genetic   analysis.
Estimates of heritability may have functional value but contribute
little to an understanding of either physiology or genetics of yield
potential.   The complexity of the processes involved inhibits an
isolation and analysis of physiological, let alone genetic components.
Here, once again, genetics may point the way, as Evans (1983)
suggests, to some of the major determinants, by the application of
molecular techniques, and, as it now appears possible, through the
use of transposable elements.   This may well open a new era in the
understanding and use of genetic resources.

I thank Dr. A.H.D. Brown for helpful discussions and for
comments on the manuscript.

REFERENCES

Brown, A.H.D., Nevo, E., Zohary, D., and Dagan, O. (1978). Genetic variation in natural populations of wild barley (Hordeum spontaneum). Genetica. 49, 97-108.

Burdon, J.J., Oates, J.D. and Marshall, D.R. (1983). Interactions between Avena and Puccinia species. I. The wild hosts : Avena barbata Pott ex Link, A. fatua L., A. ludoviciana Durien. J. Appl. Ecol. 20, in press.

Chang, T.T. (1980). The rice genetic resources program of IRRI and its impact on rice improvement, in Rice Improvement in China and other Asioan Countries, pp. 85-105. IRRI, Los Baños.

Dinoor, A. (1970). Sources of oat crown rust resistance in hexaploid and tetraploid wild oats in Israel. Can. J. Bot. 48, 153-161.

Duvick, D. (1981). Genetic diversity in major farm crops on the farm and in reserve. MS of paper given at XIII International Botanical Congress (to be published).

Evans, L.T. (1983). Raising the yield potential : by selection or design? in Genetic Engineering of Plants (Hollaender, A., Meredith, C.P. and Kosuge, T., eds.), Plenum Press, New York (in press).

Frankel, O.H. (1947). The theory of plant breeding for yield. Heredity. 1, 109-120.

Frankel, O.H. and Soulé, M.E. (1981). Conservation and Evolution. Cambridge University Press.

Frey, K.J. (1981). Capabilities and limitations of conventional plant breeding, in Genetic Engineering for Crop Improvement (Rachie, K.O. and Lyman, J.M. eds.), pp. 15-62, Rockefeller Foundation, New York.

Harlan, J.R. (1976). Genetic resources in wild relatives of crops. Crop Sci. 16, 329-333.

IBPGR-IRRI Rice Advisory Committee. (1980). Descriptors for Rice Oryza sativa L. IRRI, Manila.

Katznelson, J. and Morley, F.H.W. (1965). A taxonomic revision of Sect. calycomorphum of the genus Trifolium. I. The geocarpic species. Israel J. Bot. 14, 112-134.

Khush, G.S. and Coffman, W.R. (1977). Genetic evaluation and utilization (GEU) program. The rice improvement program of the International Rice Research Institute. Theor. Appl. Genet. 51, 97-110.

Marshall, D.R. and Brown, A.H.D. (1975). Optimum sampling
   strategies   in   genetic   conservation   in   Crop Genetic
   Resources for Today and Tomorrow   (Frankel,   O.H.   and
   Hawkes, J.G., eds.). International Biological Programme
   2, pp. 53-80, Cambridge University Press.

Marshall,   D.R.   and   Brown,   A.H.D.   (1981).   Wheat   genetic
   resources,   in   Wheat Science - Today and Tomorrow.
   (Evans,   L.T.   and   Peacock,   J.W.   eds.)   pp.   21-40.
   Cambridge University Press.
Oates, J.D., Burdon, J.J. and Brouwer, J.B. (1983). Interactions
   between Avena and Puccinia species. II. The Pathogens
   : Puccinia coronata Cds and P. graminis Pers. f. sp.
   avenae Eriks. & Henn. J. Appl. Ecol. 20, in press.

Rick, C.M. (1982). The potential of exotic germplasm for tomato
   improvement,   in   Plant Improvement and Somatic Cell
   Genetics (Vasil,   A.K.,   Scowcroft,   W.R.   and   Frey,   K.J.
   eds.) pp. 1-28. Academic Press, New York.

Rick, C.M., Fobes, J.F. and Holle, M. (1977). Genetic variation
   in Lycopersicon pimpinellifolium : evidence of evolutionary
   change in mating systems.   Plant Syst. Evol.   127,
   139-170.

Simmonds, N.W. (1983). Engineering of plants. Trop. Agric. 60,
   66-69.

GENETIC MANIPULATION OF THE MAMMALIAN GERM LINE

SEQUENCE ORGANIZATION OF THE VERTEBRATE GENOME

Giorgio Bernardi

Laboratoire de Génétique Moléculaire, Institut Jacques
Monod, 2 Place Jussieu, 75005 Paris, France.

ABSTRACT

The sequence organization of the nuclear genome of vertebrates has
been studied by density gradient centrifugation in the presence of
DNA ligands. This approach has been combined with renaturation
kinetics and with studies of specific sequences using cloned probes.
A new picture of the vertebrate genome has emerged from these
investigations.

A number of impacts of genetic manipulations on man and society
have been discussed in this Symposium. I would like to suggest that
yet another one may turn out to be most important. This is our
understanding of the organization, of the informational content, and
of the functional regulation of the human genome. At the present
time, we are still far from this goal, simply because of the for-
midable complexity of the problems involved. Yet, the technology
needed to solve them is now available and one can foresee that we
will reach, perhaps in a relatively near future, this ultimate level
of the knowledge of ourselves, understanding what determines to such
a large extent what we are,namely our own genome. When we will reach
this goal, we will fulfill the ancient Greck precept "know
yourself".

Here I would like to summarize very briefly investigations carried
out in my laboratory on this subject over the past fifteen years.
Our studies have not been limited to the human genome, because we
thought that a broader coverage of vertebrate genomes could provide
us with useful evolutionary insights.

The basic experimental approach we have used, equilibrium centrifugation of native DNA in density gradients containing DNA ligands,
has a very solid theoretical background and can be combined with both
DNA reassociation studies and recombinant DNA technology. The
resolving power of this approach depends upon differences in the
frequencies of short (oligonucleotidic) sequences in the DNA
fragments under study. This sequence-dependence, already evident in
CsCl (Wells and Blair, 1967; Corneo et al., 1968), is enhanced in
the presence of a number of DNA ligands and leads, for instance,
to a high resolution of satellite and ribosomal DNAs(Corneo et al.,
1968; Filipski et al., 1973; Cortadas et al., 1977; Macaya et al.,
1978; Meunier-Rotival et al., 1979). More interestingly, however,
we discovered that density gradient centrifugation allows the reso-

lution of mammalian and avian main-band DNAs into a small number of
discrete components, which could, therefore, be studied in detail, so
providing a wealth of new information.

Neglecting satellite and minor components, mammalian and avian
genomes can be resolved (Filipski et al., 1973; Thiery et al.,1976;
Macaya et al., 1976; Cortadas et al., 1979) in Cs_2SO_4/Ag^+ or
$Cs_2SO_4/BAMD$ (BAMD is (acetato-mercuri-methyl)dioxane), into : a) one
or two (according to the species) light components, representing
about 2/3 of the genome, and having buoyant densities in the 1.697-
1.703 g/cm^3 range; and b) two heavy components, representing about
25 % and 10 % of the genome, and having buoyant densities of 1.704
and 1.708 g/cm^3, respectively. In contrast, cold-blooded vertebrates have main-band DNAs ranging in density, in most species,
from 1.697 to 1.703 g/cm^3 (which is the range of the light components
of warm blooded DNAs) and characterized (like those of unicellular
organisms and invertebrates) by CsCl band profiles which are symme-
trical or only slightly skewed on the heavy side (Thiery et al.,
1976; Macaya et al., 1976); expectedly, DNAs from warm-blooded
vertebrates exhibit CsCl bands strongly skewed on the heavy side.
The symmetry properties of CsCl bands of vertebrate DNAs are
paralleled by their compositional heterogeneities, as assessed by
the width of CsCl bands; for instance, the compositional heterogeneities of a number of main-band DNAs of fish (Hudson et al.,1980)
are as small or smaller than those of individual major components
from warm-blooded vertebrates (Cuny et al., 1981; Olofsson and
Bernardi, 1983).

The finding of a small number of discrete major components in the
genomes of warm-blooded vertebrates raises two questions, namely
a) whether the heavy components correspond to newly formed sequences
which did not exist in cold-blooded vertebrates or to sequences
which did exist in cold-blooded vertebrates, but underwent an
increase in GC; and b) the molecular weight range in which the DNA
fragments forming the major components can be separated.

The first question is clearly answered by the finding that a number
of genes localized in the main bands of cold-blooded vertebrates
(which correspond in buoyant density to the light components of
warm-blooded vertebrates), are present in the heavy components of
warm-blooded vertebrates. Such is the case, for instance, of the
mouse α -globin gene and of the chicken α - and β -globin genes,
which are all localized on 1.708 components (Bernardi, 1979; and
paper in preparation), whereas in Xenopus they are localized on
main-band DNA fragments having a buoyant density of 1.700 g/cm^3
(G. Cuny, pers. comm.). In other words, the heavy components of
warm-blooded vertebrate DNAs are the result of regional increases in
GC, (which affect both coding and non-coding DNA sequences, at it
will be shown below), and not the result of de novo formation of
sequences.

As far as the second question is concerned, since the major components do not vary in their relative amounts nor in their buoyant

densities, whether they are obtained from DNA preparations as low as 2.10^6 or higher than 2.10^8 in molecular weight, (Macaya et al., 1976), one has to conclude that they are formed by families of fragments derived from very long chromosomal segments endowed with a remarkable compositional homogeneity, which we have called isochores. This conclusion is supported by the fact that all single-copy or clustered genes tested so far (about 20) have always been found on one single major component (a proof, by itself, of a real fractionation), even when the molecular weight of the fractionated DNA preparation was in excess of 50.10^6 (or 75 kb). Since these genes could be localized anywhere on the fragments derived by random breakage, the regions of compositional homogeneity must be in excess of 100.10^6 (or 150 kb).

Several points suggest (Cuny et al., 1981) that isochores may correspond to G- and R-bands of chromosomes : a) there are strong indications   that G-bands correspond to AT-rich, late-replicating DNA, and R-bands to GC-rich, early replicating DNA; chromosome bands would therefore differ in GC contents like iso-chores; b) G-bands are very evident in warm-blooded, but faint or absent in cold-blooded vertebrates, a feature paralleled by the very different compositional heterogeneities of DNAs from warm-blooded and cold-blooded vertebrates; c) no bands can be detected in chromosomal regions harboring an amplified gene, as expected from the fact that the amplified region is small compared to an isochore; d) the amount of DNA per chromosomal band, even taking into account high resolution banding, is more than compatible with the estimated isochore size; e) G-bands and R-bands are interspersed in chromo-somes and so are isochores, as shown by the fact that genes belonging to scattered multigene families, like the actin genes, are found on all major components of mouse and human DNAs (Soriano et al. 1982) and also at 30 different loci on metaphase chromosomes; f) G-banding patterns appear to be highly conserved in warm-blooded vertebrates, as are the relative amounts and the buoyant densities of major components as well as the localization of genes within individual major components; for example, the β-globin genes of mouse, man and rabbit are all on one light component, whereas the α-globin genes of the same species are on the heaviest component.

Concerning the problem of sequence organization, the reassociation kinetics of DNA fragments derived from isolated major components from mouse, man and chicken (Soriano et al., 1981; Olofsson and Bernardi, 1983) has shown that the relative amounts of interspersed repeated and unique sequences (as well as foldback sequences in the case of the mammalian genomes) strikingly differ in the different major components of each genome; for example, most of the inters-persed sequences of chicken are concentrated in the 1.708 component, which is the poorest in such sequences in the mouse genome. These findings lead to the general conclusion that the sequence organi-zation of mammalian and avian genomes is not uniform in different chromosomal regions and, furthermore, that it exhibits remarkable differences in different mammals and between mammals and birds. Contrary to the cliché of a universal Xenopus pattern, a wide

variety of sequence arrangements exist already within vertebrates
and within mammals. Furthermore, because of the correlation between
major components and isochores, one should consider beside the
short range (1-10 kb) distribution of repeated sequences (corresponding to the classical interspersion patterns), a long-range
(> 100 kb) distribution of repeated sequences due to the interspersion of isochores and to their different internal distribution
of repeated sequences.

A closer look at the distribution of interspersed repeats can be
obtained by using cloned repeated sequences as probes. Three families
of short repeats (Alu, B1 and CRI; Schmid and Jelinek, 1982; Stumph
et al.,1982) and two families of long repeats (Bam and Kpn; Meunier-
Rotival et al.,1982; Soriano et al., 1983)were investigated and
found to be concentrated in the heavy components (particularly the
1.708 component) and in the light components, respectively of the
DNAs of man, mouse and chicken. These results are of interest
because : a) they show a very similar if not identical distribution
for specific families of repeats, which are known to share sequence
homology; b) in no case the distribution of these families is
ubiquitous; at least one major component and often two components
"contiguous" in density are either totally free or very low in
some repeated sequences. As a consequence, the postulated mobility
of the repeats must be accompanied by strong constraints on the use
of integration sites, which can only be located in some major
components but not in others. Likewise, the integration of the
bovine leukemic proviral DNA was shown to occur at a number of sites
of the host cell genome, all of which are, however, localized in
the 1.708 component (Kettmann et al., 1979).

A final point concerns the correlation between the base composition
of repeated sequences and that of the major components in which
they are embedded. In the cases studied so far, the correlation
is excellent, the short repeats present in the heavy components
being higher in GC than the long repeats present in the light
components; in both cases, the match between the compositions of
repeats and corresponding components is remarkable (Soriano et al.,
1983). A similar correlation is found with genes and even with
gene segments (untranslated and translated regions, introns) some
of which, (the introns), are known to undergo changes by insertion
and deletion rather than by point mutation. These results agree
with the finding that, within each major component, interspersed
repeated sequences and unique sequences do not differ in their GC
contents (Soriano et al.,1981). Needless to say that the match
between coding sequences and major components in base composition
can only be obtained if constraints on codon usage exist; in fact,
these have been detected (Bernardi, 1979; and paper in preparation).
Genes located in heavy components in contrast with those located in
light components show a preference for codons rich in G and C.
Interestingly, this leads to high levels of the doublet CG,
usually avoided in vertebrate genomes, in those genes.

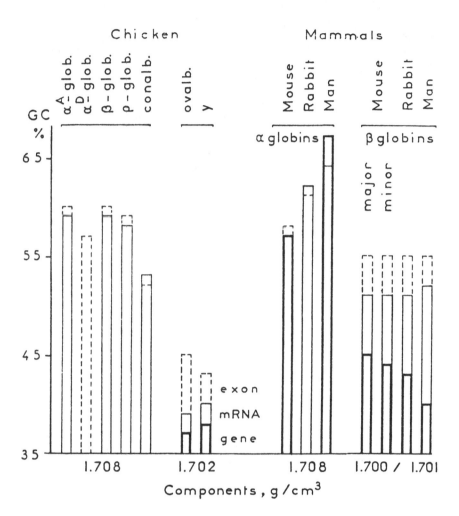

Figure 1. Histogram showing the GC content of genes (mRNAs, exons) which have been localized in the major components of chicken and mammalian genomes.

In conclusion, the discovery of isochores and the study of the
properties of DNA fragments derived from them puts the problem of
the sequence organization of vertebrate genome in a new perspective,
revealing features unsuspected so far, like the compartmentalization
of sequences into chromosomal domains, and the different sequence
organizations of genomes belonging to species relatively close to
each other from an evolutionary viewpoint. Investigations on isochores should help in bridging the gap between studies at the gene
and at the chromosomal level. Finally, the question of the origin
of isochores is raised; it is possible, that the isochores of warm-
blooded vertebrates have arisen for purely structural reasons; if
such is the case, these reasons must be really compelling, since
they appear to go as far as to impose constraints on codon usage.

REFERENCES

Bernardi, G. (1979). Organization and evolution of the eukaryotic
        genome. in Recombinant DNA and Genetic Experimentation
        (Morgan J. and Whelan W.J., eds.) pp. 15-19, Pergamon.

Corneo, G., Ginelli, E., Soave, C., and Bernardi, G. (1968).
        Isolation and characterization of mouse and guinea pig
        satellite DNA's. Biochemistry 7, 4373-4379.

Cortadas, J., Macaya, G., and Bernardi, G. (1977). An analysis of
        the bovine genome by density gradient centrifugation :
        fractionation in Cs_2SO_4/ 3,6 bis (acetato-mercuri-methyl)
        dioxane density gradient. Eur. J. Biochem. 76, 13-19.

Cortadas, J., Olofsson, B., Meunier-Rotival, M., Macaya, G.,
        and Bernardi, G. (1979).The DNA components of the
        chicken genome. Eur. J. Biochem. 99, 179-186.

Cuny, G., Soriano, P., Macaya, G., and Bernardi, G. (1981).
        The major components of the mouse and human genomes :
        preparation, basic properties, and compositional
        heterogeneity. Eur. J. Biochem. 115, 227-233.

Filipski, J., Thiery, J.P., and Bernardi, G. (1973). An analysis
        of the bovine genome by Cs_2SO_4-Ag^+ density gradient
        centrifugation. J. Mol. Biol. 80, 177-197.

Hudson, A.P., Cuny, G., Cortadas, J., Haschemeyer, A.E.V. and
        Bernardi, G. (1980). An analysis of fish genomes by
        density gradient centrifugation. Eur. J. Biochem.
        112, 203-210.

Kettmann, R., Meunier-Rotival, M., Cortadas, J., Cuny, G.,
        Ghysdaee, J., Mammericks, M., Burny, A., and Bernardi, G.
        (1979). Integration of bovine leukemia virus DNA in the
        bovine genome. Proc. Natl. Acad. Sci. U.S.A.
        76, 4822-4826.

Macaya, G., Cortadas, J., and Bernardi, G. (1978). An analysis of
    the bovine genome by density-gradient centrifugation.
    Eur. J. Biochem. 84, 179-188.

Macaya, G., Thiery, J.P., and Bernardi, G. (1976). An approach
    to the organization of eukaryotic genomes at a
    macromolecular level. J. Mol. Biol. 108, 237-254.

Meunier-Rotival, M., Cortadas, J., Macaya, G., Bernardi, G. (1979).
    Isolation and organization of calf ribosomal DNA.
    Nucleic Acids Res. 6, 2109-2123.

Meunier-Rotival, M., Soriano, P., Cuny, G., Strauss, F., and
    Bernardi, G. (1982). Sequence organization and genomic
    distribution of the major family of interspersed repeats
    of mouse DNA. Proc. Natl. Acad. Sci. U.S.A. 79, 355-359.

Schmid, C.W., and Jelinek, W.R. (1982). The Alu family of dispersed
    repetitive sequences. Science 216, 1065-1070.

Soriano, P., Macaya, G., and Bernardi, G. (1981). The major
    components of the mouse and human genomes : reassociation
    kinetics. Eur. J. Biochem. 115, 235-239.

Soriano, P., Meunier-Rotival, M., and Bernardi, G. (1983).
    The distribution of interspersed repeats is non-uniform
    and conserved in the mouse and human genomes.
    Proc. Natl. Acad. Sci. U.S.A. 80, 1816-1820.

Soriano, P., Szabo, P.and Bernardi, G. (1982).The scattered
    distribution of actin genes in the mouse and human genomes.
    EMBO J. 1, 579-583.

Stumph, W.E., Kristo, P., Tsai, M.-J., and O'Malley,.B.W. (1981).
    A chicken middle repetitive DNA sequence which shares
    homology with mammalian ubiquitous repeats.
    Nucleic acids Res. 9, 5383-5397.

Thiery, J.P., Macaya, G., and Bernardi, G. (1976). An analysis
    of eukaryotic genomes by density gradient centrifugation.
    J. Mol. Biol. 108, 219-235.

Wells, R.D. and Blair, J.E. (1967). Studies on polynucleotides.
    LXXI. Sedimentation and buoyant density studies of some
    DNA-like polymers with repeating nucleotide sequences.
    J. Mol. Biol. 27, 273-288.

DISCUSSION

A..E. SIPPEL: Did you measure C5-methylation percentages in your four
components?   I ask this because you mentioned different degrees of CG
elimination in your components.

G. BERNARDI: This was tested by comparing Hpa II and Msp I digests of
isolated components, but this method is not sensitive enough to detect
the expected changes.

N.   ZINDER: Can   you   relate   base-compositional   differences   to
chromosomes?

G. BERNARDI: All components seem to be represented on all chromosomes.

GENETIC MANIPULATION OF DROSOPHILA
WITH TRANSPOSABLE P ELEMENTS

A. Spradling, B. Wakimoto, S. Parks, J. Levine
L. Kalfayan and D. de Cicco

Department of Embryology, The Carnegie Institution
of Washington, Baltimore, Maryland 21210   U.S.A.

INTRODUCTION

Drosophila: a model eukaryote. The study of eukaryotic development comprises two broad areas: the nuclear
regulatory machinery responsible for the programming of
gene activity throughout the life cycle, and the largely
cytoplasmic mechanisms which translate these commands
into organismal physiology. The use of genetic methods,
while inextricably linked to the first, is emerging as a
valuable approach to problems in both these fields of
developmental biology. For both historical and technical
reasons, the fruitfly Drosophila melanogaster is
likely to play a major role in such studies. Intensive
studies over the last seventy have made Drosophila the
geneticaly best known eukaryote, and have established a
wealth of methods for routinely manipulating the genome
at the chromosomal level. Classical descriptive studies
of Drosophila development have not been neglected, but
perhaps most significant have been the identification of
numerous genes with interesting developmental effects.
Systematic isolation of mutations leading to phenotypes
such as female sterility, maternal effects or changes
in embryonic patterning have for the first time raised
the prospect of identifying all the loci in an organism
capable of producing a given phenotype. In many cases,
however, understanding the regulatory significance of a
gene would be aided by the systematic production and
analysis of mutations at particular sites within it. In
this review we discuss recent work in our laboratory
analyzing the developmental regulation of one family of
Drosophila genes, those involved in egg shell production,
by in vitro mutagenesis.

RESULTS AND DISCUSSION

P Element-Mediated Transformation. There are two
requirements for studying developmental properties by in
vitro mutagenesis. First the gene of interest must be
available in cloned form. In addition, methods for
efficiently introducing the mutated gene back into
a developing organism are required. Advances in the
manipulation of DNA sequences, coupled with the detailed
cytogenetic methods available in Drosophila, have made
the isolation of any Drosophila gene a relatively
routine undertaking. Recently a method for introducing
defined segments of DNA into Drosophila germline chromosomes has been developed (Spradling and Rubin, 1982;
Rubin and Spradling, 1982) as well.

The method takes advantage of the properties of a family
of Drosophila transposable elements, the P elements,
which appear to be responsible for the phenomenon of
hybrid dysgenesis (Rubin et al., 1982). Wild type
fruit flies frequently contain two varieties of P element,
"complete" elements 3kb in size and "defective" elements
which can be derived from a complete element by single
internal deletion (O'Hare and Rubin, in preparation).
Two functional properties are associated with complete
elements that appear to be useful in controlling their
activity experimentally. The first activity is an
element-coded function that is required for the mobility
of both complete and defective P elements. By analogy
with prokaryotic transposable elements it is called a
"transposase" although its molecular nature is unknown.
The second function acts to repress the transposition
P elements under most circumstances within strains
which contain them.

When DNA containing a cloned complete P element is
microinjected into the cytoplasm of an early Drosophila
embryo in the region of presumptive germ cell formation,
it is frequently incorporated into developing germ cells
where transpositions of the P element take place into
cellular chromosomes. Thus a fraction of the progeny of
an injected fly will contain the P element (but no
other sequences from the injected plasmid) as a heritable
genomic component. A strain lacking endogenous P elements
is used so that the element-encoded repressor is absent.
If plasmids containing both a complete and a defective
element are injected, transposase produced by the
complete element can catalyze the insertion of either
element into host chromosomes. Thus a gene of interest
can be introduced into Drosophila as follows. The gene
is cloned within a defective P element, mixed with a
lower concentration of DNA containing the complete
element to supply transposase, and injected into embryos.

In most of the progeny where transposition has occured, only the defective, gene-containing, P element will be present. This DNA should not undergo further transposition unless a source of transposase is introduced into the strain at a later time. A series of defective P element transformation vectors which contain multiple unique internal restriction sites have recently been developed to facilitate the construction of gene-containing transposons (Rubin and Spradling, in preparation).

Analysis of the chromosomal location of P elements following transformation demonstrated that insertion occurs at a wide variety of chromosomal sites. Thus no control over the site of integration appeared to be possible using this approach. Presumably, if sufficient chromosomal DNA surrounding a gene were included on the inserted transposon, it would function normally at any euchromatic genomic location. This is the result in the case of X-ray induced insertional translocations of large chromosome segments (see Lindsley and Grell, 1968). However Table 1 demonstrates that the efficiency of P element transposition may decrease with size, suggesting that a practical limit may exist on the total amount of DNA which can be introduced. In the absence of large regions of normal flanking sequences the expression of a gene introduced on a transposon might become highly sensitive to the position at which integration occured. Since this would make in vitro mutagenesis difficult we have investigated the effect of chromosome position on the expression of several genes following transformation.

TABLE 1. Transformation frequency with transposons containing the rosy gene.

Transposon Size(kb)	Number Injected	Larvae	Adults	Fertile	Transformed
8-10	939	284	123	85	48(5%)
13-15	895	317	181	106	18(2%)

Chromosome position. The rosy gene of Drosophila is particularly useful for studies of position effects. Only a small fraction of the rosy gene product, xanthine dehydrogenase (XDH), is necessary for the production of a visible phenotypic effect i.e. the reversion of the brown eye color characteristic of null mutants to a wild-type red color. Furthermore XDH is normally expressed in a distinctive tissue-specific manner. Only a small amount

of flanking chromosomal DNA is required for expression at
certain locations, since transposons containing an 8.2 kb
Sal I fragment confer the wild type eye color on flies
which acquire them (Rubin and Spradling, 1982).

Detailed studies of the functioning of rosy genes
introduced by transformation into rosy mutant hosts
have been carried out recently (Spradling and Rubin, in
preparation). Thirty-six isogenic lines were produced
which differ only in the location of a single heterozygous rosy transposon. Genetic and molecular studies
of these lines demonstrated that the inserts were stable,
hertible, and not subject to inactivation. Thus the
XDH activity in these lines should depend only on the
ability of the inserted gene to function at the various
chromosomal positions. We found that all the lines
expressed XDH activity in the same tissues, primarily
larval fat body and adult Malphigian tube, as a wild type
strain. Quantitative assays of XDH specific activity
showed that the genes functioned at normal levels in
most cases. At certain locations rosy genes were
somewhat more or less active; the least active gene
produced about 20% of wild type XDH activity. The most
dramatic position effects were seen when the activity
of rosy genes located on the X chromosome were compared in males and females. In eight of eight lines
tested, insertion onto the X resulted in a 40% to 100%
increase in gene activity in males compared to females.
Thus rosy genes introduced by transformation appear
to at least partially come under the influence of the
regulatory mechanisms which result in the normal doseage
compensation of X-linked Drosophila genes (see Stewart
and Merriam, 1980).

Chorion gene expression. The relative lack of position
effects on rosy gene expression suggest that the
behavior of a mutated gene reintroduced into the germline
should be interpretable. The genes which code for egg-
shell (chorion) proteins in Drosophila are particularly
favorable for study by this approach (see Mahowald and
Kambysellis, 1980). Chorion genes undergo a process of
specific gene amplification during oogenesis, and later
are expresed according to a detailed temporal program
within the ovarian follicle cells. They do not appear to
be expressed in any non-ovarian tissue. We are using
in vitro mutagenesis of these genes to determine the
DNA sequences which are necessary for these various
developmental properties.

Gene amplification. Chorion gene amplification
appears to occur by the repeated initiation of rounds
of chromosomal DNA replication at site(s) in the
vicinity of these genes (Spradling, 1981). If such

origin sequences were indeed present, they might function
at new chromosomal sites following introduction on a
P element transposon. Consequently, segments of DNA
which together comprise both clusters of the major
chorion genes have been introduced into Drosophila
chromosomes. Since such DNAs are not expected to confer
any readily detectable phenotype, the transposons used
also contained a funcional rosy gene and the host
strain was a rosy mutant. DNA was prepared from
the follicle cells of transformed strains and the
replication of the introduced chorion sequences was
determined by quantitative Southern hybridization.
These studies strongly support the idea that specific
origin sequences exist in the both gene clusters
(de Cicco and Spradling, in preparation). Most segments
of DNA showed no evidence of amplification in follicle
cells or other tissues. However fragments containing
one region of the gene cluster from chromosome III
underwent specific amplification in the follicle cells
of transformed strains. Amplification in other tissues
was not detected. As in the case of normal gene amplification, surrounding sequences were also affected. Thus
the rosy sequences adjacent to the introduced chorion
DNA, but not those at the normal rosy locus also
underwent amplification in follicle cells. The extent
of amplification of introduced sequences was about fourfold less than normally observed, however.

Temporal and tissue-specific regulation. Although
chorion genes have been the subject of substantial
genetic study, mutations which totally eliminate mRNA
production have not been isolated. In the absence of
such null mutations it would be difficult to distinguish
the product of an introduced chorion gene from the
product of the normal host genes. To circumvent this
problem, a modified gene was constructed. A 0.5kb
sequence from the phage M13 was inserted at a site near
the 5' end of the s38-1 gene. Transcription of such
a gene should result in a mRNA of increased size, and
this mRNA should hybridize to an appropriate M13 probe.
We have introduced such genes into Drosophila chromosomes by including them in P elements marked with the
rosy gene, and have studied their expression.
(Wakimoto, Kalfayan and Spradling, in preparation). Our
initial construct contained about 1kb of sequence
upstream from the 5' end. These modified genes showed
a pattern of expression indistinguishable from that of
the normal gene, although the level of expression has
not yet been determined accurately and appeared to
vary between several of the lines. mRNA complementary
to the modified gene was present only at precisely the
stages of oogenesis when normal s38-1 mRNA accumulates.
Thus normal temporal as well as tissue-specific s38-1
gene expression requires at most about 1 kb of 5'
flanking sequence.

SUMMARY

The studies reviewed here suggest that P element-mediated gene transfer provides a powerful approach to analyzing sequences required for a wide variety of developmental regulatory phenomena. The fact that transposons integrate at diverse genomic sites will probably not be a serious limitation since the expression of the genes examined was not strongly affected by chromosome position. An exception was the doseage compensation of _rosy_ genes following integration into the X chromosome.

REFERENCES

Lindsley, D. and Grell, R. (1968). Genetic variations of Drosophila melanogaster. Carnegie Inst. Washington Publ. 627.

Mahowald, A.P. and Kambysellis, M.P. (1980). Oogenesis. in The Genetics and Biology of Drosophila (Ashburner, M. and Wright, T.R.F., eds)pp. 141-224, Academic Press, New York.

Rubin, G.M. and Spradling, A.C. (1982). Genetic transformation of Drosophila with transposable element vectors. Science 218, 348-353.

Rubin, G.M., Kidwell, M.G. and Bingham, P.M. (1982). The molecular basis of P-M hybrid dysgenesis: the nature of the induce mutations. Cell 29, 987-994.

Spradling, A.C. (1981). The organization and amplifiction of two clusters of Drosophila chorion genes. Cell 27, 193-202.

Spradling, A.C. and Rubin, G.M. (1982). Transposition of cloned P elements into germline chromosomes on Drosophila melanogaster. Science 218, 341-347.

Stewart, B. and Merriam, J. (1980). Doseage compensation. in The Genetics and Biology of Drosophila (Ashburner, M. and Wright, T.R.F., eds) pp. 107-140, Academic Press, New York.

DISCUSSION

G. GROSS: Has anybody successfully used the 'P-element' system for transformation in heterologous systems, i.e. mammalian cells?

A. SPRADLING: I am aware of several attempts to transform mouse cells or embryos using P-elements; but so far at least, they were not successful.

F. GELLISSEN: Did you try to transform Drosophila cells with chorion genes derived from other insect systems?

A. SPRADLING: No.

R. JAENISCH: Are there any silent inserted genes? In other words, did you screen flies of parental phenotype for P-DNA sequences?

A. SPRADLING: Our results suggest that inactive insertions, at least those which are detectable by in situ hybridization, are rare. At least 50% of the initial transformed lines contained more than one insertion, only one of which would be required for the rosy⁺ phenotype. Yet when all 15 of the insertions from five such lines were separated and tested individually, each was capable of producing wild-type eye color. Furthermore, only a single insert was detected by hybridization in nearly 200 larvae tested from lines that behaved genetically as containing one active rosy⁺ gene.

MAKING A BIGGER MOUSE

Richard D. Palmiter[*] and Ralph L. Brinster[+]

[*]Howard Hughes Medical Institute Laboratory
Department of Biochemistry
University of Washington
Seattle, Washington
U.S.A.

[+]Laboratory of Reproductive Physiology
School of Veterinary Medicine
University of Pennsylvania
Philadelphia, Pennsylvania
U.S.A.

Animal growth is certainly under multigenic control. Although many of the genes involved are still unknown, one set is rapidly being elucidated. This set is the growth hormone (GH) cascade illustrated in Figure 1. The best known of the hormones is GH, which is synthesized in the pituitary. Its synthesis and secretion is regulated by two hypothalamic hormones: somatostatin (which is a negative regulator) and growth hormone releasing factor (which is a positive regulator). Growth hormone does not mediate growth directly, but binds to membrane receptors in the liver where it stimulates the synthesis and secretion of insulin like growth factor-1 (IGF-1), formerly called somatomedin C, which is thought to directly mediate growth by binding to membrane receptors of peripheral mesenchymal cells. Although the sequences of the proteins in this cascade pathway are known, the biochemical mechanisms by which the individual hormones act are obscure.

The genes for some of these polypeptides have been isolated, e.g. growth hormone and somatostatin, while cloning of the others is underway in several laboratories. It has been known for some time that GH from one species will interact with receptors from another. Thus, one approach to achieving larger growth of mice might be to introduce the GH gene from a compatible species into cells of a mouse by microinjecting these genes into the nuclei of fertilized eggs and then allowing these eggs to develop in the reproductive tracts of foster mothers. Although either human or rat growth hormone genes have been stably incorporated into mouse chromosomes, neither have stimulated growth, presumably because they are not expressed efficiently. If they were expressed in a tissue specific manner, i.e. only in the pituitary, it is unlikely that a few extra copies of this gene would affect growth.

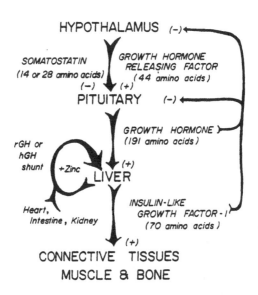

Figure 1. *Growth hormone cascade.*

These observations suggested that an alternative approach was necessary. Clearly, a strong transcriptional promoter that would function in large secretory organs, such as the liver, would amplify the amount of hormone produced tremendously. Having shown previously that the mouse metallothionein-I gene (MT-I) is expressed in a large number of tissues, including liver, kidney and intestine, and that 1 to 3 thousand MT-I mRNAs accumulate per cell in response to heavy metals such as cadmium and zinc, and knowing that this promoter would function when fused to other structural genes such as viral thymidine kinase, we constructed plasmids containing the mouse MT-I promoter fused to the structural genes of either rat or human GH (Figure 2).

Linear DNA fragments containing these genes were microinjected into the male pronuclei of fertilized mouse eggs and the animals that developed from these eggs were analysed for presence of the foreign genes in their tails by DNA dot hybridization.

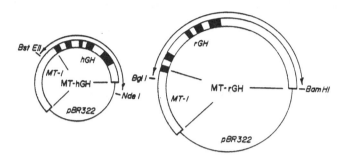

Figure 2. *Plasmids containing the mouse metallothionein-I
(MT-I) promoter fused to either human growth hormone (GH) gene
(left) or rat GH gene (right). The growth hormone exons are
shown as solid regions, introns are open and the MT-I promoter is
stippled. Linear DNA fragments indicated by the arrows were
isolated and injected into pronuclei of fertilized mouse eggs.*

A large fraction (>50%) of the animals carrying either rat or
human MT-GH fusion genes grew significantly larger (up to 2
times) than their littermates.

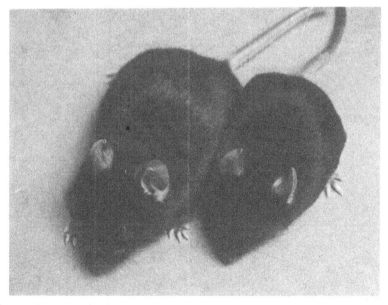

Figure 3. *A typical transgenic mouse (left) expressing a MT-rGH
fusion gene which has grown to about twice the size of a normal
littermate (right) that does not carry the fusion gene.*

This enhanced growth was correlated with a high level of GH mRNA production in the liver, intestine and heart. In some of the largest mice, about a thousand rat or human GH mRNA molecules per cell could be detected in the liver and the concentration of circulating GH was several hundred fold higher than normal (see below). These observations readily account for the rapid growth. They also raise many fascinating questions about the regulation of growth, the inheritance of this new growth property and the potential applications of this technology.

Transgenic mice carrying MT-GH fusion genes appear to start growing faster than littermates a few weeks after birth and then after about 14 weeks of age they grow at a rate more like controls (Figure 4). We suspect that GH is present in huge excess both before and after the growth spurt; thus raising the questions of what limits growth before 3 weeks and after 14 weeks. Perhaps responsiveness to GH or IGF-I are developmentally modulated. Other considerations include the role of GH and/or IGF-1 in feedback inhibition of endogenous production of mouse GH or IGF-1 (see Figure 1).

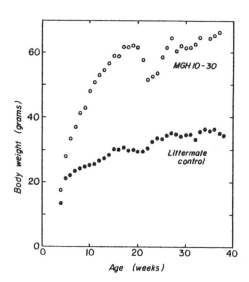

Figure 4. *Growth curve of the two mice shown in Figure 3. The dip in weight of MGH-10-30 between 20 and 25 weeks was due to an infection that was successfully treated with chloromycetin.*

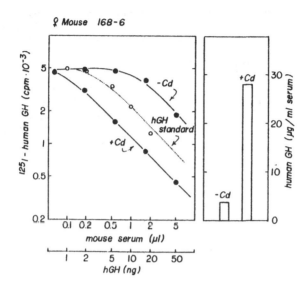

Figure 5. *Regulation of human GH production by metals. A transgenic mouse carrying MT-hGH genes was bled, then injected with $CdSO_4$ (1 mg/kg) and bled again the next day. hGH content in the sera was measured by radioimmunoassay with anti-hGH (left). The results show that Cd treatment increased the circulating hGH content about 6 fold (right). Normal mouse GH levels are about 0.1 µg/ml.*

The production of GH in these transgenic mice is regulated by Cd or Zn. We have shown a 5-10 fold induction of MT-GH gene transcription, mRNA and protein in response to acute or chronic metal stimulation. An example of the effect of Cd administration on circulating hGH levels is shown in Figure 5. In all of the animals studied thus far, growth rate was not increased by treatment with metals, presumably because the constitutive level of GH produced in the absence of metals exceeded that necessary for maximum growth. We anticipate that some animals with lower constitutive levels of MT-GH gene expression will show metal-dependent growth. Alternatively, by genetically manipulating the mouse MT-I promoter we may be able to achieve tighter control over transcription rates.

MT-I genes are expressed in most body tissues, but to different extents; thus an intriguing question is whether the foreign MT fusion genes will be expressed proportionately to the endogenous genes. The expression of high levels of hGH mRNA in some of these transgenic animals coupled with sensitive solution hybridization assays for hGH mRNA and mouse MT-I mRNA make this type of analysis possible. Figure 6A shows the standard curve used to quantitate hGH mRNA, while Figure 6B shows titration curves for hGH mRNA hybridization in various mouse tissues of a transgenic animal. Similar titration curves were obtained for mouse MT-I mRNA and the hybridization values were converted into molecules per cell.

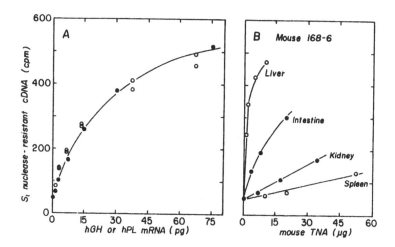

Figure 6. *Hybridization assay used to quantitate hGH mRNA levels in various mouse tissues. A ^{32}P-labeled cDNA made from a human placental lactogen (hPL) clone was used in a titration assay with human placenta (solid symbols) or human pituitary RNA (open symbols) RNA. This cDNA hybridizes equally well to either mRNA which isn't surprising considering that these genes are very homologous. This cDNA was then used to quantitate hGH mRNA levels in various mouse tissues of a transgenic mouse carrying MT-hGH fusion genes. The titration curves for several tissues are shown. The hGH mRNA content can be calculated from these curves by comparison with the standard and correcting for the RNA/DNA ratio of the individual total nucleic acid (TNA) samples. These values are shown in Table 1.*

The results from one transgenic mouse are summarized in Table 1. It is clear that hGH mRNA is expressed in all tissues examined and that there is some correspondence with mouse MT-I mRNA levels; however, the ratios are not constant. A perfect correlation would suggest that the quantitative expression of these genes is determined predominantly by the activity of diffusible regulatory molecules acting on a promoter having a permissive chromatin structure similar to that of the endogenous MT-I genes. If, however, there were significant deviations from proportional expression that were not consistently observed when comparing different animals then one might suspect that the chromosomal environment of the foreign and endogenous MT promoters has an overriding influence. Another possibility is that animals with different integration sites of the foreign genes might consistently show the same disproportionality when comparing one tissue with another, then one might think the ratio of mRNA stabilites differs from one tissue to another. This comparative analysis is being performed now. Note that it is much more demanding than asking, for example, whether foreign globin genes are expressed only in red blood cells.

Table 1. Concentration of mouse MT-I mRNA and MT-hGH mRNA in various mouse tissues of a transgenic animal (168-6) that had been stimulated with CdSO$_4$, an inducer of MT-I gene expression.

Tissue	MT-I mRNA	MT-hGH mRNA	MT-hGH mRNA / MT-I mRNA
(molecules/cell)			
Liver	2560	820	.32
Intestine	680	33	.05
Kidney	680	4.6	.01
Heart	210	21.5	.10
Brain	160	2.2	.01
Spleen	33	1.6	.05
Lung	31	1.3	.04

Numerous pedigree analyses of transgenic mice indicate that
the foreign genes are usually inherited as though they were
integrated into a single chromosome, even when there are multiple
copies of the gene. This observation is consistent with the fact
that multiple copies are generally integrated as head-to-tail
tandem arrays. A likely explanation for these tandem arrays is
that they are generated by homologous recombination among the
injected DNA molecules either before or during integration.
However, homologous integration into the mouse chromosome has not
yet been demonstrated; thus, we assume that the integration sites
are random.

Figure 7 shows that two generations of the offspring of a
transgenic mouse carrying MT-rGH fusion genes grew significantly
faster than their littermates, the average increase in adult size
being about 2-fold.

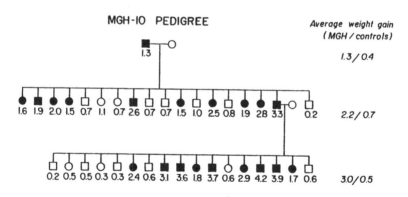

Figure 7. Inheritance of the MT-rGH and accelerated growth
rate in the MGH-10 pedigree. The MT-rGH gene is inherited by
approximately 50% of the offspring (solid symbols) as would be
expected from integration of the foreign DNA into one of the
mouse chromosomes. All of the offspring that receive the MT-rGH
gene grow significantly faster than their normal littermates.

Discussion

There is approximately a 4-fold difference in adult size when comparing various dwarf strains of mice, e.g., *little* ,that have severely depressed levels of GH and the transgenic mice described here that maintain a huge excess of circulating GH. The effects of GH deficiency or surplus are not manifest until a couple of weeks after birth, indicating (a) that other genes control fetal growth and (b) that the GH response system is developmentally regulated, becoming functional about 2-3 weeks after birth. These observations raise fundamental questions about what gene(s) control fetal growth and what limits GH response. In addition it is clear that GH response is limited to a few months, possibly due to exhaustion of all growth potential or because of homeostatic mechanisms that down-regulate the GH response system. The GH response system includes GH receptors, those factors which stimulate IGF-1 production, IGF-1 receptors and their second messages which mediate growth. Conceivably, the same step in this cascade may limit growth before and after the window of GH-responsiveness. In any event, it seems clear that the hypothalmic control of GH synthesis and secretion is not involved because in the transgenic animals, GH is being produced in extra-pituitary organs and would be insensitive to hypothalamic hormones.

In addition to providing new approaches to fundamental questions of growth control, gene transfer as described here may be useful for creating animal models of certain human genetic diseases. For example, an inbred strain of mice expressing high levels of GH would provide a convenient model of giantism in which the physiological and pathological consequences of chronic exposure to high levels of GH could be addressed. It is becoming clear, for example, that the cyclical production of GH is important in mediating many sex-dependent polymorphisms. We would predict that these sex differences would be obliterated in these transgenic animals that are chronically exposed to elevated GH.

Acknowledgments

We are indebted to our colleagues, Greg Barsh, Neal Birnberg, Howard Chen, Ronald Evans, Richard Gelinas, Robert Hammer, Gunnar Norstedt and Michael Rosenfeld for their invaluable contributions that made these studies possible. We also thank Myrna Trumbauer and Mary Yagle for their much appreciated technical assistance and Abby Dudley for typing and assembling this manuscript. The work was supported by grants from the National Institutes of Health (HD-09172 and HD-17321).

General References

Ralph L. Brinster, Howard Y. Chen, Myrna Trumbauer, Allen W. Senear, Raphael Warren and Richard D. Palmiter. Somatic Expression of Herpes Thymidine Kinase in Mice following Injection of a Fusion Gene into Eggs. *Cell* 27, 223-231, 1981.

Diane M. Durnam and Richard D. Palmiter. Transcriptional Regulation of the Mouse Metallothionein-I Gene by Heavy Metals. *J. Biol. Chem.* 256, 5712-5716, 1981.

Richard D. Palmiter, Ralph L. Brinster, Robert E. Hammer, Myrna E. Trumbauer, Michael G. Rosenfeld, Neal C. Birnberg, and Ronald M. Evans. Dramatic growth of mice that develop from eggs microinjected with metallothionein-growth hormone fusion genes. *Nature* 300, 611-615, 1982.

Richard D. Palmiter, Howard Y. Chen and Ralph L. Brinster. Differential Regulation of Metallothionein-Thymidine Kinase Fusion Genes in Transgenic Mice and Their Offspring. *Cell* 29, 701-710, 1982.

DISCUSSION

D. COOPER: Have you studied the DNA methylation of the integrated human growth hormone genes? In particular, are gene copies so modified and does this modification differ between gene copies, tissues and/or individuals?

R. PALMITER: No, we have not yet studied DNA methylation in animals with human growth hormone genes; however, we have published some experiments (in Cell 29, 701) on methylation of metallothionein-thymidine kinase fusion genes.

R. JAENISCH: Does the extent of methallothionein-thymidine induction depend on the number of DNA copies or on the basal level of expression?

R. PALMITER: There is no obvious effect of gene dose on the basal level or fold induction, but we do not know what number of genes are being expressed in those animals with multiple copies.

N. ZINDER: Can you make these insertions homozygous?

R. PALMITER: In one case we have succeeded, but in another case we have failed. In that case the heterozygous males do not transmit the foreign gene. We are investigating the possibility that the foreign gene affects spermatogenesis or is an embryonic lethal.

R. MILLER: Why is the male pronucleus a better recipient of injected DNA than the female?

R. PALMITER: Perhaps because the protamines of the sperm nucleus are exchanged for histones shortly after penetration.

K. ILLMENSEE: Have you obtained transformed mice which have integrated the injected growth hormone gene on one of the two X-chromosomes, and if so, is the integrated gene subjected to X-chromosomal inactivation or is it constitutively being expressed?

R. PALMITER: We have not yet observed integration into the X-chromosome yet, but R. Jaenisch has.

R. JAENISCH: We have derived mice carrying 20 copies of genomic retroviral copies on the X-chromosome. Transmission is strictly sex linked. So far we have no evidence for X-inactivation of these sequences.

RETROVIRUSES AND MAMMALIAN DEVELOPMENT

R. Jaenisch

Heinrich-Pette-Institut für Experimentelle Virologie und
Immunologie an der Universität Hamburg, Martinistrasse 52,
2000 Hamburg 20, Federal Republic of Germany

ABSTRACT

Retroviruses were used as model genes to study mechanisms of gene
regulation in early mammalian development. Our studies defined an
efficient de novo methylase activity which biologically inactivates
genetic information introduced into preimplantation mouse embryos.
Insertion of proviral genomes into the germ line occurred with high
frequency and fourteen different substrains of mice have been derived, each carrying the Moloney leukemia virus genome at a different Mendelian locus. Mutation of the Mov-13 locus, which was
shown to be the $\alpha 1(I)$ collagen gene by insertional mutagenesis, led
to a recessive lethal mutation and early embryonic death.

Expression of the experimentally inserted genes during later life
appears to be influenced by the chromosomal position at which integration takes place. This may be relevant for the tissue-specific
expression of developmentally regulated genes which are introduced
into mammalian embryos.

INTRODUCTION

The first successful experiment of introducing genes into animals
was achieved by microinjecting SV40 DNA into mouse blastocysts
(Jaenisch and Mintz 1974). Experimental insertion of foreign genetic
material into the germ line of mice has been accomplished by exposing early embryos to infectious retroviruses (Jaenisch 1976;
Jähner and Jaenisch 1980; Jaenisch et al. 1981) or by microinjecting
cloned DNA into zygotes (Brinster et al. 1981; Costantini and Lacy
1981; Gordon and Ruddle 1981; Harbers et al. 1981a; Wagner et al.
1981; Palmiter et al. 1982a,b; T.A. Stewart et al. 1982). While the
frequency of successful germ line integrations in such experiments
was high, expression of the inserted genes in animals derived from
manipulated embryos was unpredictable and has been observed in a few
instances only.

Retroviruses are favorable models to study gene expression in animals. The stable association of proviral DNA with the host genome is
an obligatory step in their life cycle, and as endogenous viruses
they can coexist with their host species as germ line determinants
(Jaenisch 1983). We have analyzed the interaction of retroviruses
with mouse embryos to study mechanisms of gene regulation in early
mammalian development. In this review article, I shall summarize

our results on the fate of genetic material introduced into early
embryonal cells and on the use of retroviruses as insertional mutagens to identify genes involved in embryonic development.

I. Derivation of mouse strains carrying retroviral genomes in the
germ line. By exposing early mouse embryos to Moloney leukemia virus
(M-MuLV), we have inserted the genome of this retrovirus into the
germ line of mice. Fourteen different substrains of mice were obtained, each carrying a single copy of the Moloney leukemia virus as
a Mendelian gene (Jaenisch 1976; Jähner and Jaenisch 1980; Jaenisch
et al. 1981). These substrains differ in their genotype (different
chromosomal integration sites; Mov loci) as well as in their phenotype of virus expression: the majority of substrains do not show
virus expression at all, and four substrains express virus at different stages of development. In Table 1 the characteristics and the
time of virus activation during development in the different Mov
substrains are summarized. Recent evidence obtained in our laboratory (Fiedler et al. 1982) has indeed suggested that tissue-specific
activation of viral genomes carried in the germ line of mice may be
regulated by mechanisms similar to those proposed for the tissue-
specific activation of developmentally regulated genes (Razin and
Riggs 1980). Our results suggest that the chromosomal position at
which virus integration occurs influences both the time in development of gene activation and the cell in which the proviral genome
becomes expressed (Jähner and Jaenisch 1980; Jaenisch et al. 1981;
Harbers et al. 1981a).

II. De novo methylation of retroviral genomes prevents expression in
early embryonal but not in differentiated cells. The biological importance of DNA methylation in CG nucleotides for gene expression
in eukaryotes has been established in a number of systems (Razin and
Riggs 1980; Felsenfeld and McGhee 1982). We observed a direct correlation between expression and methylation of retroviruses. All proviral genomes carried in the Mov substrains were highly methylated,
were not expressed in the tissues tested and were not infectious in
a transfection assay (Stuhlmann et al. 1981). However, when the
methyl groups were removed by molecular cloning of the proviral copies, they were rendered highly infectious (Harbers et al. 1981b;
Chumakov et al. 1982). Our results strongly suggested that DNA
methylation plays a causative role in gene regulation during development and differentiation.

The Mov substrains, with the exception of Mov-13, were derived by
exposing preimplantation mouse embryos to M-MuLV. (Mov-13 was derived from virus-exposed postimplantation embryos.) Since the infecting retroviral DNA was not methylated, de novo methylation of
the proviral genomes must have occurred at some point either during
development of the infected embryo and/or as a consequence of their
transmission through the germ line. Furthermore, it was shown previously that early mouse embryos, as well as embryonal carcinoma
(EC) cells (Jaenisch et al. 1975; Teich et al. 1977; Jaenisch 1980;
Speers et al. 1980), which have many features in common with embryonic ectoderm cells of the early mouse embryo, do not permit the
replication of M-MuLV. The experiments summarized below were done

TABLE 1. Mouse strains with germ line-integrated Moloney leukemia virus

Strain	M-MuLV sequences (genetic locus)	Expression of virus		Other characteristics
		Viremia	Time of activation	
BALB/c	Mov-1	+	1 week after birth	Virus on chromosome 6
ICR	Mov-2	(+)	In 20% as adults	
	Mov-3	+		
	Mov-4	-		Deletion in env gene
129	Mov-5	-		
	Mov-6	-		Deletion in env gene
	Mov-7	-		
	Mov-8	-		
	Mov-9	+		
	Mov-10	-		
	Mov-11	-		
	Mov-12	-		
C57BL	Mov-13	+	During embryogenesis	Gray hair
	Mov-14	+		Virus on X-chromosome

The table summarizes the characteristics of mouse sub-strains carrying M-MuLV in their germ line. For details, see Jaenisch et al. (1981).

to understand the factors that prevent expression of viral genomes introduced into early embryos and to correlate this with DNA methylation.

Two experimental approaches were used: Firstly, the fate of the infecting viral DNA was directly followed and compared in tissue culture by infecting pluripotent EC cells or differentiated cells. In a second approach, preimplantation or postimplantation mouse embryos

R. Jaenisch

were exposed to M-MuLV, and viral genomes carried in the adult animals derived from the infected embryos were characterized. In both
experimental approaches the expression of viral genomes was studied
by the XC plaque assay, quantitative RNA hybridization and/or in
situ hybridization. Modifications of the viral genomes were characterized by restriction enzyme analysis and by transfection assay of
the high molecular weight DNA. The results of these experiments have
been published (Jähner et al. 1982; C.L. Stewart et al. 1982).

The results of both the in vitro studies using EC cells and the in
vivo studies with mouse embryos led to similar conclusions and are
summarized in Table 2. Retroviral genomes introduced into early embryonal cells remained unmethylated and were expressed as long as
they remained in an episomal state (Harbers et al. 1981a) but became
de novo methylated soon after chromosomal integration (C.L. Stewart
et al. 1982). Genomic integration and de novo methylation correlated
with inhibition of gene expression. Similarly, retroviral genomes in
animals which were derived from preimplantation embryos exposed to
virus were highly methylated and not expressed (Jähner et al. 1982).
In contrast, when retroviral genomes were introduced into differentiated derivatives of EC cells or into postimplantation mouse embryos, no de novo methylation and no inhibition of gene expression
was observed upon genomic integration.

    TABLE 2. Expression and methylation of proviral genomes
introduced into cells at different stages of embryonal development

Developmental stage of virus exposure	M-MuLV genomes			
	Before integration		After integration	
	Methylated	Expressed	Methylated	Expressed
Preimplantation embryo or undifferentiated EC cells	-	+	+	-
Postimplantation embryo or differentiated EC cells	-	+	-	+

        Embryonal carcinoma (EC) cells or mouse embryos were ex-
posed to M-MuLV and analyzed as described by C.L. Stewart et al.
(1982) and Jähner et al. (1982).

The de novo methylation activity in embryonal cells appears to be of
general significance, as not only viral but also globin DNA and a
metallothionein-thymidine kinase fusion plasmid, which were microinjected into mouse zygotes, were found to be methylated in the
adult (Palmiter et al. 1982a; F. Costantini and E. Lacy, personal
communication). Our results suggest that embryonal cells possess an
efficient de novo methylase activity that inactivates any foreign
DNA which is introduced into the early embryo. Activation of the

introduced genetic material during later development may be correlated to demethylation events, which may in turn depend on the
chromosomal site where the gene was inserted.

III. Embryonic lethal mutation induced by retrovirus insertion into
the α1(I) collagen gene. Since retroviruses are capable of inserting
into many regions of the host genome, they are able to affect the
expression of cellular genes. This can occur either by the enhancing
of gene transcription from a proximal position, under the influence
of the strong viral promoter in the long terminal repeat (LTR), or
by physically disrupting cellular genes. Both phenomena have been
observed. Enhanced expression of cellular onc genes by promoter insertion (Hayward et al. 1981; Payne et al. 1981) was shown to lead
to malignant transformation. In tissue culture, Varmus et al. (1981)
induced phenotypic reversion of transformed cells by infection with
M-MuLV, and Kuff et al. (1983) demonstrated the presence of A-type
particle proviral genomes in mutated immunoglobulin genes. In animals two spontaneous mutations have been shown to be associated with
insertion of retroviral genomes into the germ line of mice (Jenkins
et al. 1981; Copeland et al. 1983). In the latter examples, however,
it is not clear at what stage of development the virus has entered
the germ line causing mutations in the respective mouse strains, nor
have the mutated genes been identified as yet.

The Mov substrains described in Table 1 were derived by exposing
mouse embryos to virus at either the preimplantation (Mov-1 to
Mov-12, Mov-14) or the postimplantation stage (Mov-13). The respective M-MuLV genomes were maintained by mating normal females
with males heterozygous for a given Mov locus [♀(wt/wt) x ♂(Mov-13/
wt)] and identifying virus-carrying offspring by testing for viremia
or for the presence of virus-specific sequences in DNA from liver
biopsies (Jaenisch et al. 1981). Heterozygosity at any of the fourteen Mov loci did not interfere with normal development or postnatal
life. The experiments summarized in Table 3 were performed to in-
vestigate whether homozygosity at the Mov loci was compatible with
normal development. Animals heterozygous for a Mov locus were mated
[♀(Mov/wt) x ♂(Mov/wt)] and the genotype of the offspring was analyzed (Jaenisch et al. 1983). From parents heterozygous for the Mov-1
to Mov-12 loci, respectively, three classes of offspring, carrying
two copies (Mov/Mov), one copy (Mov/wt) and no (wt/wt) M-MuLV-spe-
cific sequences, were obtained at approximately the expected ratio
of 1:2:1 (Table 3). This indicated that the M-MuLV genome segregated
according to Mendelian expectations and that homozygosity at the
Mov-1 to Mov-12 loci had no detectable effect on normal development.
In contrast, no homozygous offspring or embryos older than day 15 of
gestation were obtained from parents heterozygous at the Mov-13 locus (Table 3). This suggested that M-MuLV integration led to a recessive lethal mutation and embryonic arrest in this substrain. When
pregnant females at days 13 and 14 of gestation were analyzed, approximately 25% degenerated embryos were found. Genotyping of these
embryos revealed that degenerated embryos were invariably homozygous
(Mov-13/Mov-13) and that normal-appearing embryos were either heterozygous or negative for M-MuLV (ratio 1:2:1). The results suggested that integration of M-MuLV at the Mov-13 locus led to in-

sertion mutagenesis, resulting in embryonic death between day 12 and
day 13 of gestation.

TABLE 3. M-MuLV genotypes of offspring from parents heterozygous at Mov locus [♀(Mov/wt) x ♂(Mov/wt)]

	Ratio of offspring with genotype		
	Mov/Mov	Mov/wt	wt/wt
Mov-1 to Mov-12, Mov-14			
Adults	1	2	1
Mov-13			
Adults Embryos (days 15-19)	0	2	1
Embryos (days 11-14)	1^a	2	1

Offspring were derived from mice heterozygous at a given
Mov locus. DNA was isolated from individual animals or embryos
and the genotype was determined as described by Jaenisch et al.
(1983). ^{a}Homozygous embryos at days 13 and 14 were arrested in
development and dead.

To study whether the virally induced mutation in this mouse strain
affects general cell metabolism, rather than interfering with a specific stage of embryogenesis, we have prepared cell cultures from
day-12 embryos. Genotyping of cultures derived from individual embryos showed that cells homozygous at the Mov-13 locus can be grown
in tissue culture. Furthermore, using the Mov-13 locus as a probe,
we could demonstrate that this cellular gene is actively transcribed
into two mRNA species in cells from normal embryos and in cells from
embryos heterozygous at the Mov-13 locus. In cells that are homozygous at the Mov-13 locus, however, no transcripts were found. This
indicated that virus insertion interfered with transcription of this
genetic region. Further in vivo experiments showed that the Mov-13
locus became abundantly transcribed at day 12 of gestation, the time
when homozygous embryos were arrested in development. These observations suggested that insertion of M-MuLV at the Mov-13 locus led to
interruption of a gene function which is crucial for midgestation
development of the mouse but is not essential for growth of cells in
tissue culture (Schnieke et al. 1983).

To characterize tissue-specific expression of the Mov-13 locus, a
number of cell lines of mouse, rat, rabbit and human origin were
analyzed. High expression was restricted to fibroblastic, myogenic
and chondrocytic cells and was not found in hematopoietic, in endo-
derm or ectoderm derived cells. Furthermore, expression of the gene
was sensitive to transformation by sarcoma viruses. These observations suggested to us that the gene product may be secretable and
represent a protein of the extracellular matrix, and prompted us to

test the available collagen gene clones. Restriction enzyme analyses showed that the proviral genome has inserted at the 5' end of the αl(I) collagen gene, leading to a complete transcriptional block (Schnieke et al. 1983). An extensive analysis of Mov-13 embryos should allow us to understand the complex role collagen may play during mammalian embryogenesis.

CONCLUSIONS

The introduction of genetic material into early mammalian embryos is an efficient way of experimentally inserting foreign genes into the germ line and may lead to phenotypic changes in the manipulated animal. Our current lack of understanding of the molecular parameters of gene regulation in mammalian embryogenesis, however, does not allow us to predict the result of genetic manipulations, since one of the major unresolved problems remains that of tissue-specific expression of an inserted gene in the animal.

From our studies with retroviruses the following salient points emerge. Early embryonic cells possess a characteristic and efficient de novo methylation activity which modifies and biologically inactivates retroviral DNA and probably any other genetic information which is introduced into the embryo. Insertion of the introduced genes into the germ line, however, occurs with high efficiency and can lead to lethal mutations. In some instances the introduced genes become activated during later development, which probably depends on their site of chromosomal integration. If this assumption is correct, the activation of an inserted gene would depend on cis-acting regulatory mechanisms which may involve demethylation. Therefore, "correct" expression of an inserted developmentally regulated gene may depend on being integrated at a limited set of specific chromosomal sites.

ACKNOWLEDGMENTS

I thank Dr. Colin Stewart for comments on the manuscript. The work from the author's laboratory was supported by grants from the Deutsche Forschungsgemeinschaft and the Stiftung Volkswagenwerk. The Heinrich-Pette-Institut is financially supported by Freie und Hansestadt Hamburg and Bundesministerium für Jugend, Familie und Gesundheit.

REFERENCES

Brinster, R.L., Chen, H.Y., Trumbauer, M., Senear, A.W., Warren, R. and Palmiter, R.D. (1981). Somatic expression of herpes thymidine kinase in mice following injection of a fusion gene into eggs. Cell 27, 223-231.

Chumakov, I., Stuhlmann, H., Harbers, K. and Jaenisch, R. (1982). Cloning of two genetically transmitted Moloney leukemia proviral genomes: correlation between biological activity of the cloned DNA and viral genome activation in the animal. J. Virol. 42, 1088-1098.

Copeland, N.G., Jenkins, N.A. and Lee, B.K. (1983). Association of the lethal yellow (A^y) coat color mutation with an ecotropic murine leukemia virus genome. Proc. Natl. Acad. Sci. U.S.A. 80, 247-249.

Costantini, F. and Lacy, E. (1981). Introduction of a rabbit β-globin gene into the mouse germ line. Nature 294, 92-94.

Felsenfeld, G. and McGhee, J. (1982). Methylation and gene control. Nature 296, 602-603.

Fiedler, W., Nobis, P. and Jaenisch, R. (1982). Differentiation and virus expression in BALB/Mo mice: endogenous Moloney leukemia virus is not activated in hematopoietic cells. Proc. Natl. Acad. Sci. U.S.A. 79, 1874-1878.

Gordon, J.W. and Ruddle, F.H. (1981). Integration and stable germ line transmission of genes injected into mouse pronuclei. Science 214, 1244-1246.

Harbers, K., Jähner, D. and Jaenisch, R. (1981a). Microinjection of cloned retroviral genomes into mouse zygotes: integration and expression in the animal. Nature 293, 540-542.

Harbers, K., Schnieke, A., Stuhlmann, H., Jähner, D. and Jaenisch, R. (1981b). DNA methylation and gene expression: endogenous retroviral genome becomes infectious after molecular cloning. Proc. Natl. Acad. Sci. U.S.A. 78, 7609-7613.

Hayward, W.S., Neel, B.G. and Astrin, S.M. (1981). Activation of a cellular onc gene by promoter insertion in ALV-induced lymphoid leukosis. Nature 290, 475-480.

Jähner, D. and Jaenisch, R. (1980). Integration of Moloney leukemia virus into the germ line of mice: correlation between site of integration and virus activation. Nature 287, 456-458.

Jähner, D., Stuhlmann, H., Stewart, C.L., Harbers, K., Löhler, J., Simon, I. and Jaenisch, R. (1982). De novo methylation and expression of retroviral genomes during mouse embryogenesis. Nature 298, 623-628.

Jaenisch, R. (1976). Germ line integration and Mendelian transmission of the exogenous Moloney leukemia virus. Proc. Natl. Acad. Sci. U.S.A. 73, 1260-1264.

Jaenisch, R. (1980). Germ line integration and Mendelian transmission of exogenous C-type viruses, in The Molecular Biology of Tumor Viruses (Stephenson, J., ed.) pp. 131-162, Academic Press, New York.

Jaenisch, R. (1983). Endogenous retroviruses. Cell 32, 5-6.

Jaenisch, R. and Mintz, B. (1974). Simian virus 40 DNA sequences

in DNA of healthy adult mice derived from preimplantation
blastocysts injected with viral DNA. Proc. Natl. Acad.
Sci. U.S.A. 71, 1250-1254.

Jaenisch, R., Fan, H. and Croker, B. (1975). Infection of preimplantation mouse embryos and of newborn mice with leukemia
virus: tissue distribution of viral DNA and RNA and leuke-
mogenesis in the adult animal. Proc. Natl. Acad. Sci.
U.S.A. 72, 4008-4012.

Jaenisch, R., Jähner, D., Nobis, P., Simon, I., Löhler, J., Harbers,
K. and Grotkopp, D. (1981). Chromosomal position and
activation of retroviral genomes inserted into the germ
line of mice. Cell 24, 519-529.

Jaenisch, R., Harbers, K., Schnieke, A., Löhler, J., Chumakov, I.,
Jähner, D., Grotkopp, D. and Hoffmann, E. (1983). Germline
integration of Moloney murine leukemia virus at the Mov-13
locus leads to recessive lethal mutation and early embryonic
death. Cell 32, 209-216.

Jenkins, N., Copeland, N., Taylor, B. and Lee, B. (1981). Dilute
(d) coat colour mutation of DBA/2J mice is associated with
the site of integration of an ecotropic MuLV genome.
Nature 293, 370-374.

Kuff, E.L., Feenstra, A., Lueders, K., Smith, L., Hawley, R.,
Hozumi, N. and Shulman, M. (1983). Intracisternal A-par-
ticle genes as movable elements in the mouse genome. Proc.
Natl. Acad. Sci. U.S.A. 80, 1992-1996.

Palmiter, R.D., Chen, H.Y. and Brinster, R.L. (1982a). Differential
regulation of metallothionein thymidine kinase fusion genes
in transgenic mice and their offspring. Cell 29, 701-710.

Palmiter, R.D., Brinster, R.L., Hammer, R.E., Trumbauer, M.E.,
Rosenfeld, M.G., Birnberg, N.C. and Evans, R.M. (1982b).
Dramatic growth of mice that develop from eggs microinjected
with metallothionein-growth hormone fusion genes. Nature
300, 611-615.

Payne, G.S., Courtneidge, S.A., Crittenden, L.B., Fadly, A.M.,
Bishop, J.M. and Varmus, H.E. (1981). Analysis of avian
leukosis virus DNA and RNA in bursal tumors: viral gene expression is not required for maintenance of the tumor state.
Cell 23, 311-322.

Razin, A. and Riggs, A. (1980). DNA methylation and gene function.
Science 210, 604-609.

Schnieke, A., Harbers, K. and Jaenisch, R. (1983). Embryonic lethal
mutation in mice induced by retrovirus insertion into αl(I)
collagen gene. Nature, submitted.

Speers, W.C., Gautsch, J.W. and Dixon, F.J. (1980). Silent infection of murine embryonal carcinoma cells by Moloney murine leukemia virus. Virology 105, 241-244.

Stewart, C.L., Stuhlmann, H., Jähner, D. and Jaenisch, R. (1982). De novo methylation, expression, and infectivity of retroviral genomes introduced into embryonal carcinoma cells. Proc. Natl. Acad. Sci. U.S.A. 79, 4098-4102.

Stewart, T.A., Wagner, E.F. and Mintz, B. (1982). Human β-globin gene sequences injected into mouse eggs, retained in adults, and transmitted to progeny. Science 217, 1046-1048.

Stuhlmann, H., Jähner, D. and Jaenisch, R. (1981). Infectivity and methylation of retroviral genomes is correlated with expression in the animal. Cell 26, 221-232.

Teich, N.M., Weiss, R.A., Martin, G.R. and Lowy, D.R. (1977). Virus infection of murine teratocarcinoma stem cell lines. Cell 12, 973-982.

Varmus, H.E., Quintrell, N. and Ortiz, S. (1981). Retroviruses as mutagens: insertion and excision of a nontransforming provirus after expression of a resident transforming provirus. Cell 25, 23-26.

Wagner, E.F., Stewart, T. and Mintz, B. (1981). The human β-globin gene and a functional viral thymidine kinase gene in developing mice. Proc. Natl. Acad. Sci. U.S.A. 78, 5016-5020.

DISCUSSION

J. HORST: You have isolated 13 mouse strains, in which you have integrated the MoV-gene, and you have demonstrated results concerning MoV-homozygote offspring of strain 13. How do the MoV-hormozygote animals from the other strains look like?

R. JAENISCH: All other MoV-strains were derived from exposing preimplantation embryos to virus 13. MoV was derived by exposing postimplantation stage embryos. This suggests that at the early stage a different set of potential sites is available for virus integration, than at the postimplantation stage. The latter may have a higher chance of hitting an active gene.

CHROMOSOME TRANSLOCATIONS AND ONCOGENE ACTIVATION IN BURKITT LYMPHOMA

C.M. Croce

The Wistar Institute of Anatomy and Biology,
Philadelphia, Pennsylvania 19104 U.S.A.

ABSTRACT

In Burkitt lymphomas with the 8;14 chromosome translocation we have shown that the c-myc oncogene translocates from its normal position on chromosome 8 to the heavy chain locus on chromosome 14. As a result, high levels of c-myc transcripts coded for by the translocated c-myc oncogene are expressed in the tumor cells, whereas the c-myc oncogene on normal chromosome 8 is transcriptionally silent. We have also investigated Burkitt lymphoma cells with the t(2;8) and t(8;22) chromosome translocations by taking advantage of somatic cell genetic techniques to determine whether in the variant translocations similar mechanisms of c-myc activation occur. The results of these investigations indicate that in the various Burkitt lymphomas, the c-myc oncogene which is activated by the chromosomal translocation, remains on chromosome 8 and that the genes for the constant regions of kappa and lambda light chains translocate from their normal position on chromosomes 2p and 22q respectively to a chromosomal region distal (3') to the oncogene. These results indicate that activation of the c-myc gene can occur by placing the immunoglobulin gene sequences either in front (5') or behind (3') the oncogene.

INTRODUCTION

Burkitt lymphoma is a malignancy of B cells which affects predominantly children (Helne et al, 1979) Specific chromosome translocations have been observed in this disease (Manolov and Manolova 1972; Zech et al, 1976). In about 90% of cases a reciprocal translocation between chromosome 8 and 14 is observed (Manolov and Manolova 1972; Zech et al, 1976); in the remaining 10% the reciprocal translocations involve either chromosomes 2 and 8 or 8 and 22, the breakpoint on chromosome 8 being consistently on band q24 (Miyoshi et al, 1979; Bernheim et al, 1981) Since we and others have assigned the genes for immunoglobulin heavy chains and kappa and lambda light chains to chromosomes 14 (Croce et al, 1979) 2 (McBride et al, 1982; Malcolm, et al, 1982), and 22 (Erikson et al, 1981) respectively, we have speculated that the human immunoglobulin genes might be involved in the translocations observed in Burkitt lymphomas (Erikson et al, 1981).

In order to determine whether this is the case we produced somatic cell hybrids between mouse myeloma and human Burkitt lymphoma
cells with the t(8;14) translocation and examined the hybrid
cells for the presence of the chromosomes involved in this
translocation and for the presence, rearrangements, and
expression of the human immunoglobulin chain genes (Erikson et
al, 1982). The results of those studies indicate that in Burkitt
lymphomas with the t(8;14) translocation the chromosome break
occurs within the heavy chain locus and that the genes for the
variable regions of heavy chain translocate from chromosomes 14
to the involved chromosome 8 and that the expressed heavy chain
allele is located on the normal chromosome 14 (Erikson et al,
1982). We have also found that the human homologue, c-myc, of
the avian myelocytomatosis virus oncogene, v-myc, is located on
chromosome 8 and it translocates to chromosome 14 in Burkitt
lymphomas with the t(8;14) translocation (Dalla Favera, et al,
1982). Further investigations have indicated that in some of the
Burkitt lymphoma cells the c-myc gene rearranges head-to-head
with the C μ gene while in others the c-myc is translocated but is
not rearranged within a large BamH1 restriction fragment
(Dalla Favera, et al, 1982; Dalla Favera, et al, 1983). Figure 1
summarizes the previous findings concerning Burkitt lymphoma with
the t(8;14) chromosome translocation.

In the present study we have investigated the expression of the
c-myc gene in malignant and non-malignant B cells and the mechanisms of oncogene activation in Burkitt lymphomas carrying the
variant chromosome translocations.

RESULTS

Transcription of the c-myc oncogene in Burkitt lymphoma cells.
We have examined the levels of c-myc transcripts in Burkitt
lymphoma cells, in human lymphoblastoid cells and in HL60 human
promyelocytic cells which contain an amplified c-myc oncogene
which is expressed at a high level (Dalla Favera, et al, 1982a)
by S_1 protection method (Nishikura, et al, 1983). As shown in
Fig. 2, Burkitt lymphoma cell lines express high levels of c-myc
transcripts. These levels are higher than in the lymphoblastoid
cells we have examined (Fig. 2) (Nishikura, et al, 1983). As
shown in Fig. 2 the levels of c-myc transcripts of the three
Burkitt lymphoma cell lines with the t(8;14) translocation
(Daudi, CA46 and P3HR1) were even higher than in HL60 cells.

We have also examined the expression of the c-myc transcripts in
Burkitt lymphomas with head-to-head (5'to 5') rearrangements of
the c-myc gene with the Cμ gene by the Northern blotting procedure using nucleic acid probes specific for the untranslated
leader of the myc message (that is coded for by the first exon)
and for the coding segment (second and third exon) of the c-myc
oncogene (Nishikura, et al, 1983) As shown in Fig. 3B the probe
specific for the first exon of the c-myc gene does not detect the
2.3 kb myc transcripts that are expressed in Burkitt lymphoma

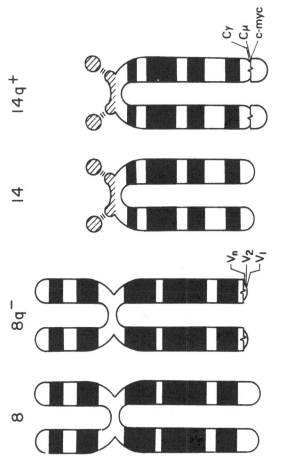

Figure 1. Diagrams of the t(8;14) chromosome translocation in Burkitt lymphoma cells. The Cμ and Cγ genes are proximal to the breakpoint on chromosome 14, while the V$_H$ gene translocates to the 8q$^-$ chromosome. The human c-myc gene on the broken chromosome 8 translocates to the heavy chain locus. Whereas in the Burkitt lymphoma cell lines, CA46 and JD38-IV, the c-myc gene is in the same 22-kb BamHI fragment with the Cμ gene, the c-myc gene is not joined to the Cμgene in P3RH-1 and Daudi Burkitt lymphoma cells.

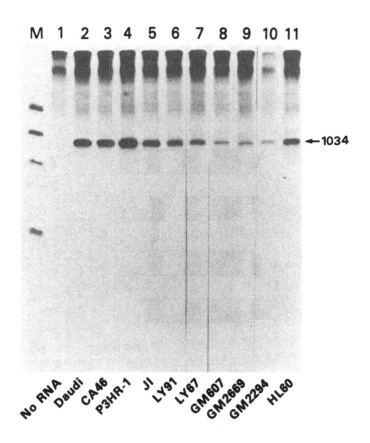

Figure 2.   Detection of the transcripts produced from the c-myc
gene in various Burkitt lymphoma cells and human lymphoblastoid
cell lines by S_1 nuclease analysis.   The probe, cleaved with BclI
and 5'-^{32}P end labeled pRyc-7.4 plasmid, was heat denatured,
hybridized in 80% formamide to 20 μg cytoplasmic RNA at 55°C,
digested with S_1 nuclease and analyzed by electrophoresis on a 7M
urea 4% polyacrylamide gel (37).   RNA from Burkitt lymphoma
cells.   Lanes 2 - 7, Daudi, CA46, P3HR-1, JI, LY91 and LY67.   RNA
from lymphoblastoid cell lines.   Lanes 8-10, GM607, GM2669 and
GM2294.   Lane 11, RNA from promyelocytic leukemia cell line.
HL-60.   Lane M, size marker: Φ X174 digested with HaeIII   and
5'-^{32}P-labeled.   The human myc RNA protect a DNA fragment 1034
nucleotide long.

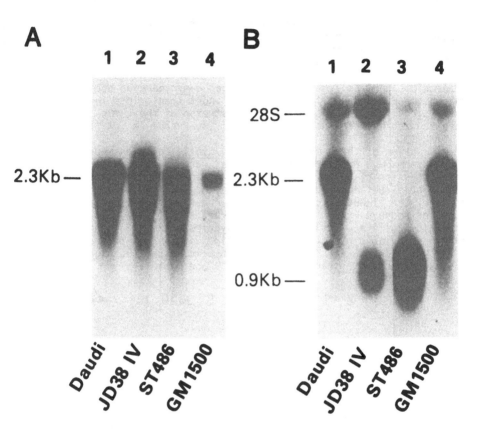

Figure 3.   Northern blotting analysis of three Burkitt lymphoma
and an Epstein-Barr virus transformed lymphoblastoid cell line
(GM1500). Cytoplasmic RNA was extracted and 20 ug of RNA were
added to each lane of 1% agarose gel. Following agarose gel
electrophoresis and Southern transfer the nitrocellulose filters
were hybridized (A) with the Ryc 7.4 probe (probe B) and (B) with
the 5' exon probe (probe A). In lanes 1, 2 and 3 the RNAs from
the lymphoma cell lines Daudi, JD38 IV and ST486 respectively.
The DNA from GM1500 cells is in lane 4. All three lymphoma cell
lines show a 2.3 kb c-myc transcript using the Ryc 7.4 probe. On
the contrary we detect the 2.3 kb myc transcripts in Daudi and
GM1500 cells using the 5' exon probe.   We detected 0.9 kb
transcripts hybridizing with the 5' exon probe in the two
lymphoma cell lines with a rearranged c-myc oncogene. The intense
hybridization of the 5' exon probe with 28 S ribosomal RNA is due
to high G/C content.

cells, in JD38 and ST486 cells, while the probe specific for the
second and third exon does (Fig. 3A). This result indicates that
JD38 and ST486 lymphoma cells with the t(8;14) translocation does
not produce normal myc transcripts and that only the rearranged
c-myc gene is transcribed, since we did not detect normal
transcripts derived from the unrearranged c-myc gene on normal
chromosome 8. In addition we detected short transcripts (~ 900
nucleotides long) derived from the first exon of the c-myc gene
involved in the translocation (Fig. 3B) (ar-Rushdi et al, 1983).

Transcription of the c-myc oncogenes in hybrids between mouse
plasmacytoma and Burkitt lymphoma cells. We have also examined
the expression of human and mouse c-myc transcripts in somatic
cell hybrids between mouse plasmacytoma and Burkitt lymphoma
cells by the S1 nuclease protection procedure using either a
human or mouse c-myc cDNA probe (Nishikura et al, 1983). As
shown in Fig. 4 and Table 1 only the hybrids with 14q+ chromosome
expressed human c-myc transcripts,hybrids with normal chromosome
8 did not (Fig. 4A and Table 1). All hybrids expressed high
levels of mouse myc transcripts (Fig. 4B and Table 1).

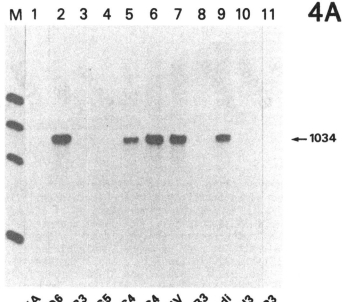

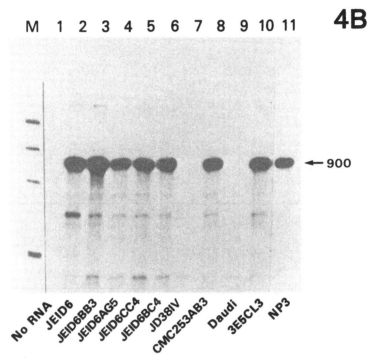

Figure 4.   S₁ nuclease analysis of c-myc RNAs in the hybrid
cells between NP3 and Burkitt lymphoma cell lines with the
t(8;14) chromosomal translocation. (A) Cytoplasmic RNA (20 ug)
was hybridized with human c-myc probe or (B) mouse c-myc S₁
probe. The parental NP3 used for hybrid preparation is a nonpro-
ducer mouse myeloma.

Therefore we conclude that only the translocated myc gene is
transcribed at a high level in Burkitt lymphoma cells and that
the juxtaposition of the myc gene and of the heavy chain locus
removes the translocated c-myc gene from its normal transcriptional control. Constitutive high level of expression of the c-
myc gene product might be responsible for the expression of
malignancy in Burkitt lymphoma.

Genetic analysis of Burkitt lymphoma cells with the variant
t(8;22) and t(2;8) chromosome translocations. We have studied
somatic cell hybrids between mouse myeloma cells and BL2 Burkitt
lymphoma cells carrying a t(8;22) chromosome translocation for
the presence (Fig. 5) and expression (Fig. 6) of human immunoglobulin chains and for c-myc oncogene. The results indicate that
the c-myc oncogene remains on the 8q⁺ chromosome, and that the
excluded and rearranged Cλ allele translocates from chromosome 22
to this chromosome 8 (Table 2) (Croce et al, 1983). As a result

Table 1

Transcription of the mouse and human c-myc genes in mouse x human hybrids

Parental cells and Hybrid clones	Human Isozymes		Human Chromosomes				Human Oncogenes		Levels of c-myc Transcripts	
	GSR	NP	8	8q⁻	14	14q⁺	c-mos	c-myc	mouse	human
P3HR-1 (BL)	+	+	+	+	+	+	+	+	-	+++
JE1D6 (NP3xP3HR-1 hybrid)	+	+	+	-	-	+	+	+	+++	+++
BB3 (NP3xP3HR-1 hybrid)	+	-	+	-	-	-	+	+	+++	-
AG5 (NP3xP3HR-1 hybrid)	+	-	+	-	-	-	+	+	+++	-
CC4 (NP3xP3HR-1 hybrid)	-	+	-	-	-	+	-	+	+++	+++
BC4 (N?3xP3HR-1 hybrid)	-	+	-	-	-	+	-	+	+++	+++
NP3 (mouse plasma-cytoma)	-	-	-	-	-	-	-	-	+++	-
JD38 (NBL)	+	+	+	+	+	+	+	+	-	+++
253 A-B3 (NP3xJD38)	+	-	+	-	+	-	+	+	+++	-
Daudi (BL)	+	+	+	+	+	-	+	+	-	+++
3E5 CL3 (NP3xDaudi hybrid)	+	+	+	+	+	-	+	+	+++	-

GSR, glutathione reductase; NP, nucleoside phosphorylase.

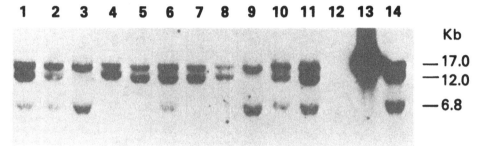

Figure 5.   Southern blotting analysis of NP3 x BL2 somatic cell
hybrids following EcoR1 digestion of cellular DNA. The nitrocellulose filters were hybridized with a Cλ genomic clone. Lanes
1-11, NP3 x BL2 hybrid DNAs. Hybrid 1-15 DNA is in lane 5;
hybrid 1-18 DNA is in lane 7 and hybrid 1-23 is in lane 8.
Hybrid 3-1 is in lane 3 and hybrid 1-9 DNA is in lane 9. NP3
mouse myeloma DNA is in lane 12, PAF DNA is in lane 13 and BL2
parental is in lane 14.

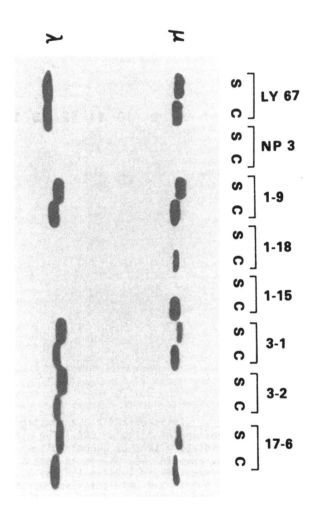

Figure 6.   Immunoprecipitation of human immunoglobulin chains produced by NP3 x BL2 hybrid clones. S: culture supernatant, C: cytosol. LY67 Burkitt lymphoma cells carry a t (8;22) translocation. Immunoprecipitation of BL2 culture supernatant or cytosol with the antihuman immunoglobulin antisera resulted in a band pattern identical to hybrids 1-9, 3-1 and 17-6 (data not shown).

Table 2. Ig genes and oncogenes in BL2 hybrids

| Cell line | Human Chromosomes* | | | | Human Cλ genes | | | Expression of human λ chains | Human oncogenes | | Transcripts of human c-myc |
	8	8q+	22	22q⁻	17 kb (upper band)	12 kb (middle band)	6.8 kb (lower band)		c-myc	c-mos	c-myc
BL2	+	+	+	+	+	+	+	+	+	+	++
NP3	-	-	-	-	-	-	-	-	-	-	-
1-9	++	-	++	+	+	-	+	+	+	+	n.d.
1-15	+|	++	-	+	+	+	-	-	+	+	+++
1-23	++	++	-	++	+	+	-	-	+	-	n.d.
3-1	++	-	++	+	+	-	+			+	+	-
3-2	+|	-	++	++	+	-	+	+	+	+	-
4-35	-	+	+	+	+	+	+	+	+	+	n.d.
17-6	-	++	++	+	+	+	+	+	+	+	+++

*Frequency of metaphases with relevant chromosome: - = none; + = <10%; +| = 10-30%; ++ = >30%.

n.d. = not done.

of the translocation, transcriptional activation of the c-myc
oncogene on the rearranged chromosome 8 (8q+) occurs (Table 2),
while the c-myc oncogene on the normal chromosome 8 is transcriptionally silent (Croce et al, 1983). These findings suggest that
the translocation of a rearranged immunoglobulin lambda locus to
the 3' side of an unrearranged c-myc oncogene (Fig. 7) may
enhance its transcription and contribute to malignant transformation (Croce et al, 1983).

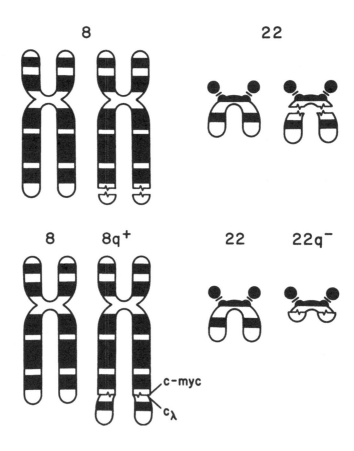

Figure 7.   Diagram of the t (8;22) translocation in Burkitt
lymphoma.   As shown in the figure the C λ locus moves from its
normal location on chromosome 22 to a region distal to the c-myc
oncogene, which is untranslocated and unrearranged in BL2 cells.

We have also studied somatic cell hybrids between mouse myeloma
and JI Burkitt lymphoma cells carrying a t(2;8) chromosome
translocation for the expression of the human κ chains and for the
presence and rearrangements of the human c-myc oncogene and the
chain genes. Our results indicate that the c-myc oncogene is
unrearranged and remains on the 8q$^+$ chromosome of JI cells (Table
3) (Erikson et al, 1983). Two rearranged C κ genes were detected:
the expressed allele on normal chromosome 2 and the excluded
allele that was translocated from chromosome 2 to the involved
chromosome 8 (8q$^+$) (Table 3). The distribution of Vκ and Cκ
genes in hybrid clones retaining different human chromosomes
indicated that Cκ is distal to the Vκ on 2p, and that the breakpoint in this Burkitt lymphoma is within the V κ region (Table 3)
(Erikson et al, 1983). High levels of transcripts of the c-myc
gene were found when it resided on the 8q$^+$ chromosome but not on
the normal chromosome 8, demonstrating that translocation of a
locus to a region distal to the c-myc oncogene enhances c-myc
transcription (Fig. 8) (Erikson et al, 1983).

CONCLUSIONS

The results described in this study indicate that in Burkitt
lymphoma with the t(8;14) translocation the c-myc oncogene
translocates from its normal position on band q24 of chromosome
14 to the heavy chain locus on chromosome 14. On the contrary in
the more infrequent Burkitt lymphomas with variant translocations
the c-myc gene remains on chromosome 8 and either the lambda or
the kappa locus translocates to a region distal to the c-myc
oncogene. Independently of the head-to-head (5' to 5') or the
head-to-tail (5' to 3') arrangement between the immunoglobulin
and the c-myc loci the c-myc oncogene on the involved chromosome
8 becomes transcriptionally highly active and is removed from
normal transcriptional control. Thus high constitutive levels of
c-myc expression occur as a result of the translocation in
Burkitt lymphoma cells. These high constitutive levels of myc
expression might have an essential role in the malignant proliferation of B cells in Burkitt lymphoma.

Table 3.   Human κ genes and oncogenes in JI x NP3 hybrids

Cells	Human chromosomes*				Human isozymes**		Human κ chains Expression	Human genes			Human Oncogenes		Human c-myc transcripts
	8	8q⁺	2	2p⁻	MDH	IDH		Cκ 15 kb	7.5 kb	Vκ	c-mos	c-myc	
JI	++	++	++	++	+	+	+	+	+	+	+	+	+++
JI 4-5	-	++	-	-	+	-	-	+	-	+	+	+	+++
JI 4-5B7	-	+	-	-	+	-	-	+	-	+	+	+	++
JI 4-5H11	-	-	-	-	-	-	-	-	-	-	-	-	-
JI 5-4	+	-	-	++	-	+	-	-	-	+	+	+	-
JI 6-5	-	-	++	-	+	+	+	-	+	+	-	-	-
JI 4-2L	-	++	-	-	+	-	-	+	-	+	+	+	+++
NP3	-	-	-	-	-	-	-	-	-	-	-	-	-

* Frequency of metaphases with relevant chromosomes: - = none; + = 10-30%; ++ = >30%.

** MDH = malate dehydrogenase; IDH = isocitrate dehydrogenase.

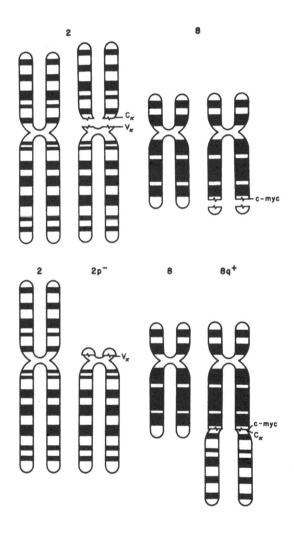

Figure 8.   Diagram of the t(2;8) translocation occurring in 5%
of Burkitt lymphomas. As shown in the figure, the V$_\kappa$ genes are
proximal and the C$_\kappa$ gene is distal on band p11 of chromosome 2.
While some of the V$_\kappa$ genes stay on the 2p⁻, the C$_\kappa$ gene translocates to the involved chromosome 8 (8q⁺).   The c-myc oncogene
remains on the involved chromosome 8 (8q⁺).

REFERENCES

ar-Rushdi, A., Nishikura, K., Erikson, J., Watt, R., Rovera, G. and Croce, C.M. (1983). Science. in press.

Bernheim, A., Berger, R. and Lenoir, G. (1981). Cancer Genet. Cytogenet. 3, 307-316.

Croce, C.M., Shander, M., Martinis, J., Cicurel, L., D'Ancona, G.G., Dolby, T.W. and Koprowski, (1979). Proc. Natl. Acad. Sci. U.S.A. 76, 3416-3419.

Croce, C.M., Thierfelder, W., Nishikura, K., Erikson, J., Finan, J., Lenoir, G., Rabbitts, T. and Nowell, P.C. (1983). Proc. Natl. Acad. Sci. U.S.A. in press.

Dalla Favera, R., Bregni, M., Erikson, J., Patterson, D., Gallo, R.C., Croce, C.M. (1982). Proc. Natl. Acad. Sci. U.S.A. 79, 7824-7827.

Dalla Favera, R., Martinotti, S., Gallo, R.C., Erikson, J. and Croce, C.M. (1983). Science 219, 963-967.

Dalla Favera, R., Wong-Staal, F. and Gallo, R.C. (1982a) Nature (London) 299, 61-63.

Erikson, J., Martinis, J. and Croce, C.M. (1981). Nature (London) 294, 173-175.

Erikson, J., Nishikura, K., ar-Rushdi, A., Finan, J., Emanuel, B., Lenoir, G., Rabbitts, T., Nowell, P.C. and Croce, C.M. (1983). Proc. Natl. Acad. Sci. U.S.A. in press.

Henle, W., Henle, G. and Lennette, E.T. (1979). Scientific American 241, 48-59.

McBride, D.W., Heiter, P.A., Hollis, G.F., Swan, D., Otey, M.C. and Leder, P. (1982). J. Exp. Med. 155, 1680-1690.

Malcolm, S., Barton, P., Murphy, C., Fergusson-Smith, M.A., Bentley, D.L. and Rabbitts, T.H. (1982). Proc. Natl. Acad. Sci. U.S.A. 79, 4957-4961.

Manolov, G. and Manolova, Y. (1972). Nature (London) 237, 33-36.

Miyoshi, I., Hiraki, S., Kimura, I., Miyamato, K. and Sato, J. (1979). Experientia 35, 742-743.

Nishikura, K., ar-Rushdi, A., Erikson, J., Watt, R., Rovera, G. and Croce, C.M. (1983). Proc. Natl. Acad. Sci. U.S.A. 80, 4822-4826.

Taub, R., Kirsch, I., Morton, C., Lenoir, G., Swan, D., Tronick, S., Aaronson, S. and Leder, P. (1982). Proc. Natl. Acad. Sci. U.S.A. 79, 7837-7841.

Zech, L., Haglund, V., Nilsson, N. and Klein, G. (1976). Int. J. Cancer 17, 47-56.

DISCUSSION

J. HORST: Did you find expression of c—myc or c—mos genes also in solid Burkitt lymphoma tumors, and if yes what percentage of the tumors have you investigated?

C. CROCE: We found expression of c—myc but not c—mos. We examined all the tumors.

ONCOGENIC TRANSFORMATION ACTIVATES CELLULAR GENES

Peter W.J. Rigby, Paul M. Brickell, David S. Latchman,
David Murphy, Karl-Heinz Westphal and Michael R.D. Scott*

Cancer Research Campaign, Eukaryotic Molecular Genetics
Research Group, Department of Biochemistry, Imperial
College of Science and Technology, London SW7 2AZ,
England.

*Present address: Department of Microbiology and Immunology,
University of California Medical School, San Francisco,
California 94143, U.S.A.

ABSTRACT

We have used molecular hybridization and cDNA cloning techniques to
isolate several clones which correspond to cellular mRNAs present
at a higher level in the SV40-transformed mouse cell line SV3T3 C138
than in the parental line, Balb/c 3T3. The cDNA clones of Set 1
define at least two transcription units the activation of which is a
general feature of both *in vitro* transformation and *in vivo*
tumorigenesis in the mouse. In contrast, the transcription unit(s)
defined by the cDNA clones of Set 2 is activated only in some
transformed cell lines, which have distinct biological properties,
and in a variety of hematopoietic and lymphoid tumors. We discuss
the possible roles played by these genes in oncogenic transformation
and the mechanisms which might be involved in their activation.

INTRODUCTION

The molecular changes which accompany the oncogenic transformation
of cultured animal cells have been the subject of intense study
during the last decade. Much of this work has concentrated upon
transformation by tumor viruses and the oncogenes carried by such
viruses have been extensively characterized (Tooze, 1981; Weiss
et al., 1982). More recent work has led to the discovery of
cellular genetic determinants capable of transferring the
transformed phenotype from tumors of murine and human origin to the
NIH3T3 line of cultured mouse fibroblasts (Cooper, 1982; Wigler, in
this Symposium). In many cases the transferred cellular gene is a
member of the *ras* family, originally defined by the fact that two
members of this family, Ha-*ras* and Ki-*ras*, are the oncogenes carried
by the Harvey and Kirsten strains of murine sarcoma virus (Der *et al.*,
1982; Parada *et al.*, 1982; Hall *et al.*, 1983; Wigler, in this
symposium).

Viral and cellular oncogenes are usually detected by *in vitro*
transformation assays using established lines of cultured cells. In
such assays transformation occurs as the result of a single event in
contrast to the multiple events thought to be involved in *in vivo*
tumorigenesis. Moreover, in the majority of cases transformation
is both initiated and maintained by a single gene and thus a single
protein. Only in the case of polyoma virus is there clear evidence

for the concerted action of two viral gene products during
transformation (Rassoulzadegan *et al.*, 1982).

However, the biological and biochemical changes which distinguish a
transformed cell from its untransformed parent are myriad and it
seemed to us to be intrinsically unlikely that all of these changes
could result from a direct action of the transforming protein. We
therefore decided to search for cellular genes which are
transcriptionally activated in transformed cells in the hope of
defining transcription units which respond,.directly or indirectly,
to the transforming protein and which contribute to the final
phenotype of the cell.

The transformation of mouse fibroblasts by Simian virus 40 (SV40)
provides a very promising system in which to study this problem.
The transforming protein of SV40, large T-antigen, is a complex
multifunctional protein with several properties consistent with it
having a direct role in the regulation of cellular gene expression
(Rigby and Lane, 1983). Large T-antigen binds with high affinity to
the origin of viral DNA replication which overlaps the promoter for
the early transcription unit and thus the protein can autoregulate
its own synthesis (Hansen *et al.*, 1981; Rio and Tjian, 1983). Large
T-antigen also binds to cellular DNA (Oren *et al.*, 1980) and before
we began this work there were two lines of evidence which indicated
that large T-antigen could affect the expression of cellular genes.
Following the infection of either permissive or non-permissive cells
SV40 causes the induction of a number of cellular enzymes involved in
nucleotide metabolism and DNA replication (Tooze, 1973) and in the
case of thymidine kinase there is evidence that a function of large
T-antigen is required (Postel and Levine, 1976). It is likely that
this induction is mediated at the level of transcription but there
is no proof of this. In murine-human somatic cell hybrids only one
of the two rDNA complements is expressed; introduction of large
T-antigen into such cells causes the transcriptional activation of
the previously quiescent rDNA complement (Soprano *et al.*, 1979,
1981). Learned *et al.* (1983) have recently shown that purified
large T-antigen will activate the transcription of rDNA in an *in
vitro* system. Moreover, Williams *et al.* (1977) showed, by using
mRNA/cDNA hybridization in solution, that there are detectable but
subtle changes in the cytoplasmic mRNA population following
transformation by SV40.

We therefore decided to attempt to develop techniques which would
allow us to clone cellular genes activated by SV40 transformation
in order to undertake a detailed study of the nature of such genes
and of the mechanism(s) by which they are activated.

        RESULTS

We have constructed large, plasmid-based cDNA libraries
representative of the cytoplasmic, polyadenylated mRNA populations
of normal Balb/c 3T3 cells and of an SV40-transformed derivative,
SV3T3 C138 (Rigby *et al.*, 1980). We used the high density technique
of Hanahan and Meselson (1980) to screen these libraries with

labeled mRNA from the two cell types and could detect no differences. Such colony screening procedures can detect clones homologous to mRNAs of an abundance of 0.1% or above and thus our data show, in agreement with those of Williams *et al.* (1977), that transformation does not cause significant changes in relatively abundant mRNAs. This result is somewhat surprising in view of the quite dramatic biological changes which accompany transformation.

In order to isolate cDNA clones homologous to mRNAs present at a higher level in SV3T3 C138 than in Balb/c 3T3 it was necessary to enrich the probe for such differentially expressed sequences. This was done by covalently coupling the Balb/c 3T3 cDNA library to cellulose and then repeatedly hybridizing to this ^{32}P-labeled cDNA made against SV3T3 C138 mRNA. This procedure effected a significant purification of the transformed cell specific sequences which at this stage could be clearly detected by solution hybridization (Scott *et al.*, 1983a, b). The desired sequences were further enriched by a preparative solution hybridization and the ^{32}P-labeled cDNA thus obtained was used to screen the SV3T3 C138 cDNA library. We thus isolated forty-two clones which were initially divided into four sets on the basis of restriction endonuclease fingerprinting and cross-hybridization experiments (Scott *et al.*, 1983b). We have subsequently defined a fifth set by virtue of the fact that these cDNA clones, like those of Set 1 (Scott *et al.*, 1983b), contain a sequence repeated in the mouse genome. The basic characteristics of the clones of these five sets are summarized in Table 1.

We have so far concentrated our attention on analyses of Sets 1 and 2. pAG64, the prototype of Set 1, hybridizes to three mRNAs, of 1.6Kb, 0.7Kb and 0.6Kb, which are present at a higher level in SV3T3 C138 than in Balb/c 3T3 (Scott *et al.*, 1983b). The larger (1.6Kb) and the two smaller (0.7/0.6Kb) Set 1 mRNAs are the products of two separate transcription units which are related only by the

TABLE 1.   Characteristics of the five sets of transformed cell specific cDNA clones.

Designation	Set 1	Set 2	Set 3	Set 4	Set 5
Prototype	pAG64	pAG59	pAG82	pAG88	pAG10
Insert Length	1.55Kb	1.8Kb	1.58Kb	0.70Kb	1.2Kb
pAG clones related by cross-hybridization	15, 22, 38, 64, 71, 85, 86, 104.	37, 57, 58, 59, 69.	1, 31, 75, 82.	13, 47, 48, 77, 88, 97, 98, 105, 109.	10, 41, 74.
Total number of clones in set	8	5	4	9	3

The prototype clone of each set is that with the longest insert.

presence of the dispersed repetitive element (unpublished data). The
Set 1 mRNAs are activated in all SV40-transformed mouse cell lines
that we have tested and in lines of mouse fibroblasts transformed by
another papovavirus, polyoma, by the C-type retroviruses Rous
sarcoma virus and Abelson murine leukemia virus and by the chemical
carcinogens methylcholanthrene and methylcholanthrene epoxide (Scott
et al., 1983b). We have furthermore shown that Set 1 mRNAs are also
present at high levels in a variety of hematopoietic and lymphoid
tumors, in a neuroblastoma and in a hepatoma (unpublished data).
These results are summarized in Table 2.

TABLE 2.   Expression of the Set 1 and Set 2 transcripts
in mouse transformed cell and tumor cell lines.

	Set 1		Set 2	
Cell Line	1.6Kb	0.7/0.6Kb	7.5/6.5Kb	3.2Kb
Balb/c 3T3	−	−	−	−
SV3T3 C120	+	+/−	−	−
SV3T3 C126	+	+	++	++
SV3T3 C138	+	+	+	+
SV3T3 C149	+	+/−	−	−
SV3T3 C1H	+	+	−	−
SV3T3 C1M	+	+/−	−	−
Py3T3	+	+	−	−
Py*tsa*3T3	+	−	−	−
RSV3T3 TK3 BXB4	+	+/−	−	+
RSV3T3 BC6	+	+/−	−	−
ANN-1	+	+	+	−
Balb/c AMuLV A1R1	+	−	+	+
Balb/c 1OME HD A5R1	+	+	−	−
Balb/c 1OCr MC A2R1	+	+	−	−
AKR thymoma, BW5147.A11	+	+	++	++
Mastocytoma, P815	+++	+	++	++
Friend erythroleukemia, GM86	++	+	+	+
Myeloma, SP2/O Ag14	+	+	+	+
AMuLV-induced B cell tumor	+++	−	+	+
Neuroblastoma, PLATT	+	+	+	+
Rat hepatoma, FAZA	+	+	−	−

pAG59, the prototype cDNA clone of Set 2, hybridizes to four SV3T3
C138 mRNAs, of 7.6Kb, 6.5Kb, 3.2Kb and 1.8Kb. The 3.2Kb mRNA is the
predominant species and it and the two larger transcripts are
transformed cell specific; the 1.8Kb mRNA is of very low abundance
and is not regulated by transformation (Scott *et al.*, 1983b).
Activation of the Set 2 transcription unit(s) is not a general
feature of SV40 transformation. Set 2 activation is confined to
cell lines with distinctive biological properties, for example high
tumorigenicity, and abundant expression of the 6.5Kb mRNA appears
to be associated with the acquisition of the anchorage independent
phenotype (Scott *et al.*, 1983b). Activation of Set 2 is also

restricted in mouse cells transformed by other agents but again there
is a correlation between Set 2 expression and an extreme phenotype
(Scott *et al.*, 1983b). We have also detected elevated levels of the
Set 2 mRNAs in lymphoid and hematopoietic tumors. These data are
summarized in Table 2.

DISCUSSION

We have shown that it is possible to use molecular hybridization and
cDNA cloning techniques to isolate mouse genes when the only assay
available is that the corresponding mRNAs are present at a higher
level in one cell type than in another, closely related, cell type
and that this technology can be successfully applied to genes
encoding low abundance mRNAs (Scott *et al.*, 1983a, b). We have
applied these procedures to one specific problem but it is clear
that such methods could be applied to other transformation systems
and that, in particular, they will find extensive application in
developmental biology.

The genes that we have isolated thus far have a number of very
interesting properties. Of particular importance is the fact that
the Set 1 mRNAs are present at elevated levels in all of the *in vitro*
transformed cell lines and *in vivo* tumors of mouse origin that we
have examined (Scott *et al.*, 1983b and unpublished data). Moreover,
we have preliminary evidence that the activation of these genes is
an early event during *in vitro* transformation. We are presently
isolating the homologous human clones in order to ascertain whether
activation of the Set 1 genes is a general feature of human tumor
cells. If so, and if Set 1 activation occurs early, during the
preneoplastic phase, then suitable probes, capable of detecting
either the mRNAs or their protein products, may have diagnostic value.
While Set 2 gene activation is not a general feature of oncogenesis
the fact that this gene(s) is activated in cells with particular
phenotypes means that it may be possible to use Set 2 probes to
further dissect the mechanism of transformation. We are currently
using DNA transfection techniques to introduce the Set 2
transcription unit(s), under the control of strong eukaryotic
promoters, into cultured cells in order to determine whether
elevated Set 2 expression can induce morphological transformation or
some particular aspect of the phenotype.

The other striking property of our clones is the presence within the
Set 1 transcription units of a dispersed repetitive element. We
have recently shown that although this repeated sequence is not, by
definition, present in the Set 2 cDNA clones it is contained within
the corresponding genomic clone although in this case it appears to
be outside the transcription unit. It is extremely unlikely that
the association of the Set 1 repeat with another gene activated by
oncogenic transformation is fortuitous. We believe it likely that
the repeated sequence is involved in the control of the expression
of the transformation activated genes and are presently testing this
possibility experimentally. A model for the control of embryonic
gene expression by such dispersed repetitive elements has been
proposed by Davidson and Posakony (1982).

P.W.J. Rigby *et al.*

ACKNOWLEDGEMENTS

P.M.B. and D.M. are supported by a Training Fellowship and a Research Studentship, respectively, from the Medical Research Council. K-H.W. holds a Long Term Fellowship from the European Molecular Biology Organization while P.W.J.R. holds a Career Development Award from the Cancer Research Campaign which also paid for this work. We are grateful to Sue Hayman for her expert preparation of the manuscript.

REFERENCES

Cooper, G.M. (1982). Cellular transforming genes. Science 218, 801-806.

Davidson, E.H. and Posakony, J.W. (1982). Repetitive sequence transcripts in development. Nature 297, 633-635.

Der, C.J., Krontiris, T.G. and Cooper, G.M. (1982). Transforming genes of human bladder and lung carcinoma cell lines are homologous to the *ras* genes of Harvey and Kirsten sarcoma viruses. Proc. Natl. Acad. Sci. U.S.A. 79, 3637-3640.

Hall, A., Marshall, C.J., Spurr, N.K. and Weiss, R.A. (1983). Identification of transforming gene in two human sarcoma cell lines as a new member of the *ras* gene family located on chromosome 1. Nature 303, 396-400.

Hanahan, D. and Meselson, M. (1980). Plasmid screening at high colony density. Gene 10, 63-67.

Hansen, U., Tenen, D.G., Livingston, D.M. and Sharp, P.A. (1981). T antigen repression of SV40 early transcription from two promoters. Cell 27, 603-612.

Learned, R.M., Smale, S.T., Haltiner, M.M. and Tjian, R. (1983). Regulation of human ribosomal RNA transcription. Proc. Natl. Acad. Sci. U.S.A. 80, 3558-3562.

Oren, M., Winocour, E. and Prives, C. (1980). Differential affinities of simian virus 40 large tumor antigen for DNA. Proc. Natl. Acad. Sci. U.S.A. 77, 220-224.

Parada, L.F., Tabin, C.J., Shih, C. and Weinberg, R.A. (1982). Human EJ bladder carcinoma oncogene is homologue of Harvey sarcoma virus *ras* gene. Nature 297, 474-478.

Postel, E.H. and Levine, A.J. (1976). The requirement of Simian virus 40 gene A product for the stimulation of cellular thymidine kinase activity after viral infection. Virology 73, 206-215.

Oncogenic transformation activates cellular genes 233

Rassoulzadegan, M., Cowie, A., Carr, A., Glaichenhaus, N., Kamen, R. and Cuzin, F. (1982). The roles of individual polyoma virus early proteins in oncogenic transformation. Nature 300, 713–718.

Rigby, P.W.J. and Lane, D.P. (1983). Structure and function of the Simian virus 40 large T-antigen. Adv. in Viral Oncology 3, 31–57.

Rigby, P.W.J., Chia, W., Clayton, C.E. and Lovett, M. (1980). The structure and expression of the integrated viral DNA in mouse cells transformed by Simian virus 40. Proc. Roy. Soc. Lond. Ser. B. 210, 437–450.

Rio, D. and Tjian, R. (1983). SV40 T antigen binding site mutations that affect autoregulation. Cell 32, 1227–1240.

Scott, M.R.D., Brickell, P.M., Latchman, D.S., Murphy, D., Westphal, K–H. and Rigby, P.W.J. (1983a). The use of cDNA cloning techniques to isolate genes activated in tumour cells. Haematology and Blood Transfusion 28, Modern Trends in Human Leukemia V, 236–240.

Scott, M.R.D., Westphal, K–H. and Rigby, P.W.J. (1983b). Activation of mouse genes in transformed cells. Cell, in press.

Soprano, K.J., Dev, V.G., Croce, C.M. and Baserga, R. (1979). Reactivation of silent rRNA genes by Simian virus 40 in human–mouse hybrid cells. Proc. Natl. Acad. Sci. U.S.A. 76, 3885–3889.

Soprano, K.J., Jonak, G.J., Galanti, N., Floros, J. and Baserga, R. (1981). Identification of an SV40 DNA sequence related to the reactivation of silent rRNA genes in human>mouse hybrid cells. Virology 109, 127–136.

Tooze, J. (1973). The Molecular Biology of Tumour Viruses. 1st Edition. Cold Spring Harbor Laboratory, New York.

Tooze, J. (1981). Molecular Biology of Tumor Viruses. Part 2. DNA Tumor Viruses. Revised 2nd Edition. Cold Spring Harbor Laboratory, New York.

Weiss, R.A., Teich, N.M., Varmus, H. and Coffin, J.M. (1982). Molecular Biology of Tumor Viruses. Part 3, RNA Tumor Viruses. Second Edition. Cold Spring Harbor Laboratory, New York.

Wigler, M. (1983). Human transforming genes. In Genetic Manipulation: Impact on Man and Society (Starlinger, P. and Illmensee, K., eds). ICSU Press, Miami.

Williams, J.G., Hoffman, R. and Penman, S. (1977). The extensive
    homology between mRNA sequences of normal and SV40-
    transformed human fibroblasts. Cell 11, 901-907.

DISCUSSION

A.E. SIPPEL: Determining the sequence of cloned Sel 2 and Sel 1 cDNA,
did you find open reading frames on the sequence?

P. RIGBY: The answer to that question in one case is "yes", in the
other it is slightly embarrasing. For the small RNA there is a frame
there but it is interesting that it only encodes for a protein of about
40 amino acids. You can speculate about what that might be; our
speculation is that it is a gene for a growth factor, which would make
a lot of sense in terms of what we now think about laboratory
transformation. The sequence of the 1.6Kb RNA is presently something
of an embarrassment. That pulls out the 37K protein but we have a stop
codon in the middle. There is something wrong with that sequence that
has to be checked out.

PROBLEMS OF GENETIC ENGINEERING IN ANIMAL BREEDING

G.F. Stranzinger

Institute for Animal Production, Animal Breeding Section,
ETH Zuerich Zentrum, CH-8092 Zuerich, Switzerland

ABSTRACT

Remarks are given in respect to the interaction between society, environment and animal production. Some of the major problems in connection with genetic engineering and animal breeding are listed. Specific attention is given to differences in the knowledge of the genom
between the different organisms involved. Several examples of application of genetic engineering to animal breeding are presented. Application of the technique will be especially influenced by the future economic situation.

INTRODUCTION

Plants, wild and domesticated animals, and humans live in a dynamic
equilibrium. Any disturbance of this equilibrium creats ecological
problems. In fact, due to the limited availability of energy and natural resources, human activities very frequently cause such disturb-
ancies. Such an activity is the common practice of using animals for
production of proteins and other products like hides, furs or silk.
As an example, one can mention the discrepancy between theory and
practice of animal welfare, pollution of the environment due to animal waste products, and management of intensive animal production.
However, as long as humans desire animal products and use animals for
their pleasure, concessions and further developments in animal production are essential. The use of animals includes their slaughter
and the replacement of animals for breeding and production. For an
efficient replacement, one wants to assure that the quality of the
new generation is similar, or better, to make the investment successful. To achieve this, one must adopt efficient forms of housing,
feeding regimes, selection procedures and biotechnical methods such
as artificial insemination, embryo transfer or germ cell conservation. Genetic engineering would introduce another option to the realm
of animal breeding, making it more efficient, at least in situations
requiring genetic changes and corrections. More knowledge, however
must be collected concerning the genom of farm animals before those
techniques can be used. Only efficient, exact, predictable and eco-
nomical new techniques can complement conventional methods, otherwise
techniques such as genetic engineering will be limited only to special cases.

PROBLEMS FOR GENETIC ENGINEERING IN ANIMAL BREEDING

Information from different research disciplines can not be fully applied to farm animal production and breeding procedures. The reasons
are as follows:
- Philosophical attitude of human beings concerning the value of animals, especially farm animals
- Large numbers of different farm animals are kept and produced in
  differing environments for many reasons
- Differences in the complexity of genetic interactions, due to the
  construction of the genom, specific traits in question and their
  influence to the environment
- The size, number and value of the investigated laboratory species
  can not be compared to that of farm animals
- The influence of genetic improvement of farm animals on society is
  dependent upon the success rate and productivity
- Ethical and moral aspects of the application of genetic research
  from lower organisms, laboratory animals, farm animals and humans
  are differently accepted and tolerated by the public.

Before a decision is made concerning the practical use of genetic engineering, one should fully understand the impact of the problems
mentioned above on man and society. For this purpose, the techniques
of genetic engineering should be tested even in extreme situations
when a high risk of finantial loss and a conflict with legal and moral notions may be incured.
For the sake of simplicity and on the ground of science, research
with lower organisms and laboratory animals provides much information
which can be easily and accurately proven due to the extensive knowledge of the construction of the genom in these organisms. This is
not the case for the genom of farm animals as indicated by Fries et
al. (1982) in a summary of the present state of knowledge. For example, in the pig there are only three assigned loci on three different autosomal chromosomes (2n = 38) in comparison to more than
400 loci on all chromosomes (2n = 46) in the human (McKusick, 1980).
On the other hand, due to the general breeding practice of farm animals and breeding research, there is much specific knowledge of genetic mechanisms and their interactions with the environment. This information, however, can not be extrapolated to lower organisms and
humans, and one must consider that breeding systems do not exist in
humans.
The most significant impact on man was brought about by the combination of biotechnical developments and population genetics for selection of farm animals which caused a large increase in production of
farm products in most industrialized nations. Subsequently, it has
become necessary that political influence be used to control production, despite the fact that many areas of the world are not capable
of producing sufficient farm products. The distribution of excess
products to areas of starvation is very much limited for reasons
which are not discussed here. The introduction and improvement of
conventional methods of animal breeding in these areas should be the

first step; genetic engineering can only be practiced afterwards.
Presently, the following applications of genetic engineering to animal production can be envisaged:

- According to Seidel and Amann (1982), the sale of the sperm separated into male and female components for cattle production (artificial insemination) would gross $ 50 millions per year in the USA.
- The recovery of oocytes from slaughtered female cattle would yield approximately 30 oocytes per ovary (Suess and Madison, 1983) and would provide material for recipients of cell and gene transfer, and for additional experimental designs for which other sources of female germ cell material and/or embryos are not available.
- The production of chimeras by the microinjection of cellular components from farm animals with valuable genetic traits into embryos in early stages is feasible. For example, chimeric animals with 4 haplotypes for the histocompatibility complex would be superior in an adverse environment (Stranzinger, 1983).
- Sexing of early embryos in combination with the production of chimeras could reduce fertility problems in those effected animals.
- The exploitation of techniques developed through genetic engineering (such as restriction enzymes) and used in gene mapping (Ruddle, 1981) would provide information concerning the inheritance of valuable traits and would open up the possibility for gene transfer.
- Cloning of rare superior animals with somatic cells can be accomplished by using split embryos. A part of the embryo (several cells of the omnipotent ectodermal tissue) can be conserved in liquid nitrogen for years while the other parts are transferred to recipient foster animals. The fetuses develop into adult animals which are tested conventionally for production traits. After the testing period, and in the case of superiority, the conserved twin cells can be cloned to produce at least several duplicates as described by Illmensee et al., 1982).

CONCLUSIONS

The importance of the described techniques and applications for animal breeding is obvious. Additional basic research in farm animal genetics and related areas is recommended. Efficient conventional methods of animal breeding which are on hand will continue to be used, but techniques of genetic engineering can be introduced and applied if they are economically feasible. Presently, the number of trained investigators and financial support for applied research in animal breeding, with respect to genetic engineering, are limited and should be increased before further progress can be made in this field.

ACKNOWLEDGMENTS

I would like to thank my coworkers for many suggestions and help and especially Miss V. Madison for translation.

238   G.F. Stranzinger

REFERENCES

Fries, R., Dolf, G. and Stranzinger, G. (1982). Gene mapping in farm
    animals. Present state, own investigations and possible ap-
    plications of gene maps. Schweiz. Landw. Monatshefte 60,
    205 - 221.

Illmensee, K., Bürki, K., Hope, P.C. and Ullrich, A. (1981). Nuclear
    and gene transplantation in the mouse, in Development Bio-
    logy Using Purified Genes. ICN-UCLA Symposia on Molecular
    Biology (Brown, D.D. and Fox, C.F.). vol 23 pp. 607 - 619,
    Academic Press, New York.

McKusick, V.A. (1980). The anatomy of the human genom. J. Hered. 71,
    370 - 391.

Ruddle, F.H. (1981). A new era in mammalian gene mapping. Somatic cell
    genetics and recombinant DNA methodologies. Nature 294 (5837),
    115 - 120.

Seidel, G.E. and Amann, R.P. (1982). The impact of using sexed semen,
    in Prospects for Sexing Mammalian Sperm (Amann, R.P. and
    Seidel, G.E.) Colorado Assoc. Press. Boulder, Colorado 80309.

Stranzinger, G.F. (1983). Gentechnik in der Tierzucht - heutiger Stand
    und Forschungsrichtungen. Vierteljahresschrift Schweiz.
    Naturforsch. Gesellsch. 128/3 Zuerich. in press.

Suess, U. and Madison, V. (1983). Morphology and Meiosis of Bovine
    oocytes taken from ovaries collected after slaughter. 15th
    Annual USGEB/USSBE Meeting. Fribourg, Switzerland.

                            DISCUSSION

W. VAN DEN DAELE: How do you proceed with embryos you do not transfer?
Would you consider passing them to scientists for making experiments?

G.F. STRANZINGER: Those embryos that are not consider good for
transfer are allowed to continue to grow in culture.

IN VITRO FERTILIZATION AND EMBRYOTRANSFER:
IMPLICATIONS FOR GENETIC ENGINEERING.

T.L.A. Hünlich,   S. Trotnow   and   T. Kniewald

Klinik für Frauenheilkunde mit Poliklinik und
Hebammenschule der Universität Erlangen-Nürnberg
(Director: Professor K.G. Ober, MD)
Universitätsstraße 21/23,   D-8520 Erlangen,
Federal Republic of Germany

ABSTRACT

By exposing the ovum, zygote and cleaving embryo to in vitro conditions during in vitro fertilization, the techniques of genetic engineering can be applied to omnipotent human cells.
The techniques of in vitro fertilization and embryo transfer in man
are described and world-wide results are reported.
The main problems during the course of in vitro fertilization and
embryo transfer, and the possible future developments in this field are
discussed, with respect to the application of genetic engineering.

INTRODUCTION

The consequences of in vitro fertilization and embryo transfer techniques for genetic engineering can be summarized in a few words: For
successful in vitro fertilization and embryo transfer it is necessary
to expose the ovum, the zygote and the cleaving embryo to in vitro
conditions. This provides the preconditions for any extracorporeal
genetic manipulation of the omnipotent embryonal cell.
It would be very interesting at this point, to give a review of the
historical development leading to the in vitro fertilization and embryo transfer techniques, as they are used in treatment of human sterility today. This survey would have to cover almost one hundred years,
going back to 1890, when Heape succeeded in transferring rabbit embryos to foster mothers (Heape 1890). Many other honourable names
would have to be mentioned. It is worthwhile, however, to remember at
least Whitten's work, who created the first chemically defined medium
for embryo cultures in 1956 (Whitten 1956). After this basic achieve-
ment it was Chang, who first successfully fertilized mammal ova in
1959 (Chang 1959). In his work he used the rabbit as a model.
The first human pregnancy that resulted from in vitro fertilization
was established in 1976 by Steptoe and Edwards (Steptoe and Edwards
1976). Unfortunately, it was an ectopic tubal pregnancy and had to be
treated surgically. In 1978 Luise Brown, the first so called "test-
tube-baby" was born in England (Steptoe and Edwards 1978). A rapid
development then followed, and by 1982 this new method of treatment of
human sterility was successfully offered in seven countries.

METHODS

Before a sterile couple is taken into an in vitro fertilization program,   they have to go through basic diagnostic examinations that pre-
cede any sterility treatment.   There are two essential preconditions
that have to be met before the patient will be accepted:
1)   The woman has to have at least one functioning ovary in a position suitable for follicular puncture.
2)   The woman has to have a normally developed uterus.
In the Erlangen in vitro fertilization program at this time there are
two other preconditions for acceptance:
1)   The woman's menstrual cycle has to be regular and free of endo-
crinological disturbances.
2)   Tha patient's husband has to prove his fertility by a normal
sperm count.
Therefore, in Erlangen, at the moment only women suffering from either
tubal or idiopathic sterility will be treated.   In the case of the
latter the reason for sterility is still unknown.

Ovarian stimulation.   In order to get as many mature oocytes as possible at a determined time, patients are either treated by oral medi-
cation with antiestrogens (clomiphene citrate) or intramuscular in-
jections of gonadotropins.   This treatment will stimulate several
ovarian follicles to develop simultaneously,   so that in most cases
more than one oocyte can be recovered.   We prefer the first mentioned
method.   Depending on the length of the menstrual cycle, the anti-
estrogene is given from day four or five after the onset of the last
menstrual period for five consecutive days.   The follicular growth
then is monitored daily by two parameters: firstly by measuring the
rising bloodlevels of estradiol, secondly by following the increasing
follicular diameter by ultrasound examinations.
As soon as these two parameters indicate an almost mature follicle,
we inject intramuscularily 5.000 I.U. HCG.   About 36 hrs later the
ovulation will take place.   The follicles have to be punctured before
that time   (Fig. 1).

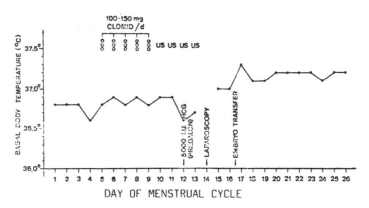

DAY OF MENSTRUAL CYCLE

Figure 1.   Time schedule for ovarian stimulation by
clomiphene citrate / HCG treatment, ultrasound monitoring (US), follicular puncture and embryo transfer.

The first pregnancies after in vitro fertilization were achieved by
utilizing spontaneous cycles (Edwards et al. 1980, Lopata et al.
1980). The hormonal treatment was believed to be disadvantageous
(Edwards et al. 1980). In these cases one had to detect the LH-surge
by monitoring the urinary or blood levels of the luteinizing hormone.
Approximately 28 hrs after this surge ovulation is to be expected in
normal human cycles. Today the LH monitoring is only employed in
patients that are difficult to be monitored by estradiol levels and
ultrasound examinations. Thus, unsuccessful attempts of follicular
puncture can be prevented.

Follicular puncture.   32 to 34 hrs after the HCG was administered,
the follicular puncture has to take place. Up to now this can be done
in three different ways: During laparotomy, laparoscopy and by employing an ultrasonically guided puncture technique, as it is successfully used in Göteborg, Sweden (Wikland et al. 1982).
We mostly puncture during a laparoscopy. With the help of long forceps, inserted through the abdominal wall in a trocar, the ovary is
put in position under visual control. Follicles are punctured with a
long cannula also running through a trocar positioned in the abdominal
wall. The follicular contents are then aspirated by means of a pump
and trapped in a tube. The negative pressure of the system can be
controlled with a pedal by the laparascopist's foot. Immediately
after recovery, the follicular fluid is taken into the laboratory,
where it is searched for the ovum.
The follicular puncture during laparotomy and the ultrasonically
guided technique are especially indicated in cases with severe pelvic
adhesions.

Insemination and embryo culture.   Depending on the maturity of the
ovum, as estimated from its morphological appearance, the husband is
asked to provide a sperm sample either immediately or some hours after
the puncture. The ejaculate is washed in medium and diluted. The
sperm suspension then is added to the ovum at a concentration of approximately 100.000 motile sperms / ml.
Culture takes place in a desiccator placed inside an incubator. We
use a humidified gas atmosphere of 5 % CO_2, 5 % O_2 and 90 % N_2.
The temperature is continuously kept at 37°C. For culture purposes as
well as for ova collection and sperm preparation we use Ham's F-10 medium of an osmolality of 280 mosmol and pH of 7.2 to 7.3. We add heat
inactivated serum obtained from umbilical cord blood to the final solution.
The culture will be checked for the first time about 20 hrs after insemination, when the pronuclei stage is to be expected. At this stage
of development one often can recognize polyploidy, and exclude the concerned cell from the further procedure. Within the next 60 hrs the
embryo developes into the eight cell stage.

Embryo transfer.   Until today embryo transfer was successful, if performed between the one- and the sixteen cell stage. Again there are
several possible methods for embryo transfer. The main difference,
however, is, whether two instruments, a guiding device and a transfer
catheter, or only one instrument is used. We employ the two-instru-
ment-method (Trotnow et al. 1982a). After the patient is in position
on a regular chair for gynecological examinations, the cervix is

grasped and stabilized with forceps by the help of a speculum. The
guiding device of the catheter is then introduced and positioned with
its tip slightly over the inner cervical os. The transfer catheter,
loaded with the embryos in less than 10 µl of medium is then led
through the guiding device, so that its tip is positioned inside the
uterine cavity. The embryos are then released, after the guiding
device has been drawn back into the cervical channel. Then both instruments are removed.

Even using as little as 3 µl of transfer medium, could not prevent an
ectopic tubal pregnancy in one of our patients (Trotnow et al. 1982b).
We therefore started blocking damaged tubes with clips during the
laparoscopy previous to the transfer.

In cases of a successful treatment, the basal body temperature stays
on a high level, the HCG-radioreceptor assay will be positive from day
12 after the puncture and the urinary assay of HCG will be positive
three weeks after the conception.

RESULTS

Last year in July, at a meeting in Murnau, West-Germany, the results
of all successful teams, dealing with in vitro fertilization and
embryo transfer were put together in a table. Figures given then can
today at least be doubled.

TABLE 1.   Pregnancies achieved by extracorporeal fertilization

	total number of children born	total number of pregnancies achieved (clinical !)	average number of embryo transfers per month
Cambridge-Bourn Hall	35 - 40	120	ca. 40
Monash Univ. Melbourne	25	60 - 65	25 - 30
Royal Women's Hospital Melbourne	5	37	ca. 20
Norfolk, Virg. USA	2	24	5 - 8 (?)
Erlangen	3	9	ca. 15
Paris / Clamart	1	10	ca. 25
Paris / Necker	-	3	ca. 12
Wien	-	12	15 - 20
Kiel	-	2	ca. 10
Göteborg	-	3	2 - 3

Data for this table were collected by Prof.H.M.Beier, Dept.
of Anatomy and Reproductive Biology, University of Aachen and Prof.
H.R.Lindner, Dept. of Hormone Research, Weizmann Institute of Science,
Revohot, Israel, by questioning participants of the Organon Symposion
"Fertilization of the Human Egg in vitro - Biological Basis and Clini-
cal Applications", Murnau, West-Germany, June 2 - 4, 1982.

Concerning the statistics of Erlangen, there are now four boys and two girls that have been born healthily. Two pregnancies were ectopic, and three ended in miscarriages. Ten more pregnancies between month three and eight have developed well until now.
In addition to the groups mentioned in table 1, there are now other successful teams in Lübeck (West-Germany), Basle (Switzerland) and Zagreb (Yugoslawia).

DISCUSSION

Main problems during in vitro fertilization and embryo transfer in man
There are several problems one is faced with, when employing the tech-niques of extracorporal fertilization in treatment of human sterility.
1)   The hormonal treatment results in individually different reac-tions of the patients, so that the follicular puncture does not always seem to take place at the most appropriate time.
2)   There are cases, in which the puncture of the ovary is not possi-ble, for instance, when it is wrapped in adhesions and positioned be-hind the uterus.
3)   The biggest problem now is probably connected with embryo trans-fer: in our statistics in vitro fertilization in man yields a success-rate of approximately 90 % of all ova that were rated mature. Only about 10 % of the embryo transfer will lead to ongoing pregnancy.

To explain this, one has to know that in vivo a certain number of em-bryos is defect and unable to implant. The estimations concerning natural conceptions, ending in early miscarriage, lie between 30 and 60 %. During in vitro fertilization this number of embryos, carrying microdefects, might even be greater.
Further it is not known, if embryos are possibly expelled through the cervix after transfer, as it is known in the case of the rabbit (Adams 1980). Whether prostaglandines play a role in this respect also is unknown.
An important aspect results from our observation that during in vitro culture some embryos tend to grow slower than others. This could lead to an asynchronism between the maternal system and the embryonal state and - the slower the embryo is growing - to a lower implantation rate.
Certainly an atraumatic, non bloody transfer is necessary for optimal success rates. Fibrin released by a trauma could possibly prevent im-plantation. There is a lot of work left to be done in this field.

Future aspects.   In connection with the main subject of this meeting future aspects of in vitro fertilization and embryo transfer have to be touched, especially in view of genetic manipulations.
The donation of ova was already tried in Australia by Trounson and his team. They transferred a heterologeously conceived embryo (sperm pa-rental) to the woman of a couple, in whom they were not able to col-lect ova. This treatment is discussed for cases of genetic risks on the side of the woman, or if the ovaries cannot be punctured.
In our hospital at Erlangen, there will be no heterologeous exchange of either ova or sperms for ethical, medical and forensic reasons.
Cryopreservation, also up to now unsuccessfully tried by Trounson, Australia (Trounson et al 1981), might be one possible way to over-come the mentioned asynchronism between the maternal system and embryo

development. Further, embryos could be saved, if transfer is not possible at the planned time because of unpredictible events.
The right procedure for freezing and thawing human embryos will be
very hard to find, on account of the small number of embryos available
for experiments. It is known that there are differences between
freezing conditions for embryos of various species, for instance the
mouse and the rat. Therefore, the knowledge of animal experiments
cannot be fully transferred to the work with human embryos. Besides
this, experiments with human embryos would be ethically questionable.
Thus, until today, there is no embryobank, even if some investigators
are thinking about it aloud.
Further future projects will be the treatment of male subfertility
and the simplification of the method itself. One might be able to
offer the treatment one day on an outpatient basis. The ultrasonically guided puncture technique is one step in this direction.
In vitro fertilization and embryo transfer during the last year have
certainly brought about a lot of knowledge concerning the handling of
human embryos in vitro. In this way various extracorporeal manipulations have been put closer to practicability. This knowledge, however, will always have to be used by physicians, biologists, lab
technicians, and also the patients. It will be their task to decide
in every case new if, and how the various techniques shall be used.

To employ genetic technologies in human embryos, however, it is not
necessary to use in vitro fertilization. Easier procedures, as for
instance tubal flushing techniques, would suffice, if one only aims
at genetic manipulations, as for instance the transfer of genes.
It is possible that these techniques could one day be used in treatment of hereditary diseases, provided there will be no greater damage
induced by the manipulation itself.
The vision of producing identical twins from "omnipotent" blastomeres
deriving from a single embryo, as reported possible in the case of
the animal by Stranzinger (in this Symposium), can be projected on
human conditions, too. Especially, as nature seems to yield identical
twins by exactly the same procedure.

REFERENCES

Adams, L.E. (1980). Retention and development of eggs transferred to the uterus at various times after ovulation in the rabbit. J. Reprod. Fert. 60, 309-315.

Chang, M.C. (1959). Fertilization of rabbit ova in vitro. Nature 184, 466-467.

Edwards, R.G., Steptoe, P.C. and Purdy, J.M. (1980). Establishing full-term human pregnancies using cleaving embryos grown in vitro. Br. J. Obstet. Gynecol. 87, 737-756

Heape, W. (1890). Preliminary note on the transplantation and growth of mammalian ova within a uterine foster mother. Proc. Roy. Soc. 48, 457-458.

Lopata, A., Johnoton, J.W.H., Hoult, I.J. and Speirs, A.J. (1980). Pregnancy following intrauterine implantation of an embryo Obtained by in vitro fertilization of a preovulatory egg. Fertil. Steril. 33, 117-120.

Steptoe, P.C. and Edwards, R.G. (1976). Reimplantation of a human embryo with subsequent tubal pregnancy. Lancet 1, 880.

Steptoe, P.C. and Edwards, R.G. (1978). Birth after the reimplantation of a human embryo. Lancet 2, 366.

Stranzinger, D.G. (1983). Problems of genetic engineering in animal breeding.

Trotnow, S., Al-Hasani, S., Kniewald, T., Becker, H. and Hünlich,T. (1982). Human embryo transfer, in Fertilization of the human egg in vitro: biological basis and clinical applications. (Beier, H.M., Lindner, H.R., eds), Springer Verlag Berlin (in press)

Trotnow, S., Al-Hasani, S., Hünlich, T. and Schill, W.B. (1982). Bilateral tubal pregnancy following in vitro fertilization and embryo transfer. Arch. Gynecol. (in press)

Trounson, A.O., Mohr, L.R., Pugh, P.A., Leeton, J.F. and Wood, E.C. (1981). The deep-freezing of human embryos. In Proceedings of III world congress of human reproduction, Berlin, 1981. Amsterdam: Excerpta Medica, 367 (Abstract)

Whitten, W.K. (1956). Culture of tubal mouse ova. Nature 177, 96.

Wikland, M., Nilsson, L. and Hamberger, L. (1982). The use of ultrasound in a human in vitro fertilization program. Abstract for the III. Meeting of the World Federation for Ultrasound in Medicine and Biology, July 26-30, 1982, Brighton, England.

DISCUSSION

NOYER-WEIDNER: Do you feel that our society really needs in vitro fertilization and embryo transfer and expensive research to overcome the problems connected with these techniques in view of the fact that thousands of children have no parents and would perhaps be rather happy to be adopted and in view of the fact that one of the major problems of mankind is overpopulation?

T.L.A. HUNLICH: There are two questions here. First, those patients whom we treat by in vitro fertilization procedures, shouldn't they not instead adopt a child? You should realize that adoption, in Germany at least, is very difficult. Many parents would be happy if they could easily adopt a child.   There are not enough children available for adoption.

Plus you know that the adoption of a child who is several years old is easier than the adoption of a real infant. I know from conversations with our patients that they really longed for their genetically-owned children and this is what we are providing by our techniques. Secondly, with regard to overpopulation and the fact that our procedures may be increasing the population I would say that if you are aiming to avoid overpopulation you should also stop treating cancer.

Subject Index